# Optics
# Essentials

# Optics Essentials

## An Interdisciplinary Guide

### Araz Yacoubian

Ler Technologies
Carlsbad, California, USA

CRC Press
Taylor & Francis Group
Boca Raton London New York

CRC Press is an imprint of the
Taylor & Francis Group, an **informa** business

CRC Press
Taylor & Francis Group
6000 Broken Sound Parkway NW, Suite 300
Boca Raton, FL 33487-2742

Printed on acid-free paper
Version Date: 20141023

International Standard Book Number-13: 978-1-4665-5117-6 (Hardback)

**Visit the Taylor & Francis Web site at**
**http://www.taylorandfrancis.com**

**and the CRC Press Web site at**
**http://www.crcpress.com**

# Contents

# Preface

The inspiration for writing this book came from working with so many engineers with great technical expertise who so often are exposed to optics but have not had any significant education on the topic. I realized there is a great need for a book that explains optics in basic terms. Many of these professionals are faced with tasks and challenges in their field that interrelate with optical components and systems, and understanding basic principles would benefit them in tackling these challenges. The aim of this book is to provide enough material to enable readers to gain a basic understanding of optics without having to commit to extended periods of study. The book is a companion that readers can use in their day-to-day work without getting distracted from their daily tasks. The more advanced topics will only be briefly discussed, and the reader will be directed to the appropriate texts for more detailed study.

Additionally, this book includes MATLAB® simulations and suggested experiments. The simulations are easily executed in MATLAB prompt (or by simply pushing the Run button), and will enable the reader to understand optical parameters and how changes affect the total outcome (e.g., how changing the lens focal length will change image position).

The experiments are designed to help readers obtain a thorough understanding of optical principles. Most experiments are done with materials and components that are readily available, and can be set up very quickly without any special tools and apparatus. For those who do not plan to perform these experiments, it is highly suggested to at least read them, because many of the optical phenomena are explained through experimental examples. After all, a "thought experiment" (plato.stanford. edu/entries/thought-experiment) can be as effective a learning tool as an actual one.

The combination of MATLAB simulations and suggested experiments will serve a broad audience, including those who are either inclined toward design and simulations or those who like more hands-on work.

## INTENDED AUDIENCE

This book is intended for the following audience

- *Industry professionals*, including electrical and electronics engineers, mechanical engineers, biomedical engineers, chemists or chemical engineers, and other technical professionals that design or use components and systems that involve optics.
- *Managers* who supervise projects that involve optics or optical hardware, or need to collaborate with other teams with optics background.
- *Researchers* with minimal or no knowledge of optics but have other physics, science, engineering, or other technical background, and plan to use optics in their respective fields.

- *Professors and researchers* who are in science or engineering fields, with minimal or no optics background, who like to either collaborate with optics faculty and professionals, or would like to learn basic optics to add functionality to their existing research projects.
- *Students* who are in electrical engineering, electronics, mechanical, biomedical engineering, chemistry, or other science and engineering fields who are just entering or planning to enter the optics field, or need to use optics in their field of study.

## HOW TO USE THIS BOOK

The choice of how to use this book depends on the reader's needs and learning preferences. The following are four potential methods that can be used to get the most of this book:

Method 1—One option is to start with the section on applications (Chapter 9, Section 9.3), find what is most relevant, and refer to that particular chapter that describes the relevant phenomenon.

Method 2—Another method is to browse through each chapter, find what seems to be most relevant, and start with that chapter.

Method 3—Start with light reading, and do the simulations and experiments that might be interesting. This will help readers identify the location of relevant information and refer back to that chapter as the need arises.

Method 4—Finally, simply read each chapter sequentially.

This book is organized to be independent of the audience's reading style. It does not require readers to commit to one style or another, yet it can be used both as a reference book as well as a basic learning tool. It is my hope to facilitate the learning process and to make learning optics an enjoyable experience.

## ACKNOWLEDGMENTS

I'd like to thank Le Nguyen Binh, Paul LoVecchio, Wen Wang, Steve Bond, and Tomasz Jagielinski for their suggestions and discussions during preparation of the manuscript. I'd also like to thank Ashley Gasque at CRC Press for her suggestions and guidance. Finally, I'd like to thank my family for their extreme patience, support, and encouragement, which made this book a reality.

**Araz Yacoubian**

# Author

**Araz Yacoubian** is a research scientist based in southern California with general experience in the optics-related fields.

He received a Ph.D. in electrical engineering-electrophysics from University of Southern California. His industry experience includes: integrated optics, nonlinear optics, optical image processing, holography, electro-optics, optical communication, and a variety of optical sensors, imagers, and components.

He worked in both research and development as well as in production environments. Throughout his career he worked on numerous optical-related projects. Examples of his work include demonstration of the first water immersible hologram and its application to contact lens tinting, development of a deployable (briefcase sized) real-time liquid crystal light valve-based image-processing unit, development of a surface and subsurface detection and imaging system based on a combination of optical/acoustic methods, and development of novel broad spectral band detectors and imagers, among others.

During his career, Dr. Yacoubian worked with many technical personnel, including electrical and mechanical engineers, chemists, and technical managers, who are faced with tasks and challenges in their perspective fields which interrelate to optical components and systems. This book, *Optics Essentials: An Interdisciplinary Guide*, was inspired while working with these professionals, with the desire to provide a valuable reference to understanding basic optical principals and to help the interdisciplinary professional tackle optics-related challenges.

# 1 Optical Systems and Components

The basic principles of optics revolve around generation, propagation, manipulation, and detection of light. Most optical systems use one or more of these principles.

We are surrounded by optical systems and components, both natural and man-made. Almost all optical systems are made up of few basic building blocks. These are

1. A light source (e.g., sunlight, flashlight, or a laser)
2. A medium where light propagates, such as air, water, or glass
3. A medium or components to manipulate light, such a as a lens to focus it, a shutter or a "light switch" for communication, or a fiber-optic cable to guide light
4. A detector, such as the cones and rods on the retina, a photodetector array used in a digital camera, or a light-sensitive film for analog photography

Additionally, optical systems frequently employ mechanical apparatus and electronics.

For example, the human eye manipulates light using a lens that focuses light and images from the scene onto a biological photodetector array (retina). The detector converts light to a signal that is sent to the brain for processing and interpreting.

A camera, for example, acts similarly to the eye, with the addition of a light source (a flash lamp) for taking photographs in low-light conditions.

Similarly, fiber optic communication systems incorporate a light source, such as a laser, a modulator to introduce variation to the light intensity (manipulation), a fiber optic cable that transmits light in a particular direction, and a detector that detects light, and sends the signal to associated electronics for interpreting the signal.

Examples of optical systems are shown in Table 1.1.

Figure 1.1 shows a how components fit into an optical system, either for optical communication or optical sensing. For example, in a fiber optic communication system, light emitted from a source (such as a laser) gets modulated with an external modulator, or the source itself (such as a laser diode) is directly modulated. It is coupled to an optical fiber that then transmits light to a distance. It is then detected with a photodetector, and the signal is demodulated and interpreted by electronics.

In nature, sunlight hits an object, is scattered, and the direction and amplitude of scattered light is determined by the shape, color, and absorption of the object. Light is then transmitted through space and is received by the observer's eye. Here the source is externally "amplitude modulated" by the object. Another example of communication/sensing is light emission from a distance star (source) that travels through space and is detected by the human eye or a telescope or camera.

**TABLE 1.1**

**Examples of Optical Systems**

| System Description | Light Source | Manipulation/ Modulation | Propagating Median | Light Collection | Detection |
|---|---|---|---|---|---|
| Human visual system | Sunlight | Reflection/ Scattering from objects, intensity variation, angular variation | Air | Lens | Rods and cones on the retina |
| Camera | Flashlight, room light, sunlight, etc. | Refraction from a lens | Air | Lens | Film (analog camera) or detector array (digital camera) |
| Fiber optic communication | Laser diode | External modulator or direct modulation of laser diode intensity | Fiber optic cable | Fiber optic coupler | Photodiode |
| Flatbed scanner or fax machine | Light array | Scattering of light from a paper | Air | Lens array | Photodetector array |

Often optical communications and optical sensing are presented as two different disciplines, but often optical systems incorporate both, as communication is enabled by sensors and detectors, and optical sensing requires communicating what is being sensed.

The science of optics revolves around everything that has to do with light. The definition of "light" has evolved throughout history, as technology to generate and detect light has evolved and the understanding of light has evolved. Today, optics, or the study of light, covers a much wider range than the visible spectrum. It often covers ultraviolet (UV), x-ray, and shorter wavelengths, and infrared and beyond. Many of the principles, however, are applicable to these ranges, and familiarizing in one region, such as in the visible spectrum, will greatly benefit the readers' ability to expand his or her knowledge into other parts of the spectrum, such as in infrared (IR).

## 1.1   CHAPTER DESCRIPTIONS

Many of the chapters describe portions that comprise an optical system, such as light source and light detection.

Chapter 2 describes light sources, such as lasers, light-emitting diodes, and thermal sources. To enable the reader to compare various light sources, photometric and radiometric parameters are discussed. Radiometric quantities are physical quantities (such as optical power measured in watts). When dealing with light sources that cover the visible spectrum, often these quantities are presented in photometric units, which are quantities that reference to the response of the human eye. Many optical systems that cover both the visible and non-visible portions of the spectrum (e.g., UV or IR)

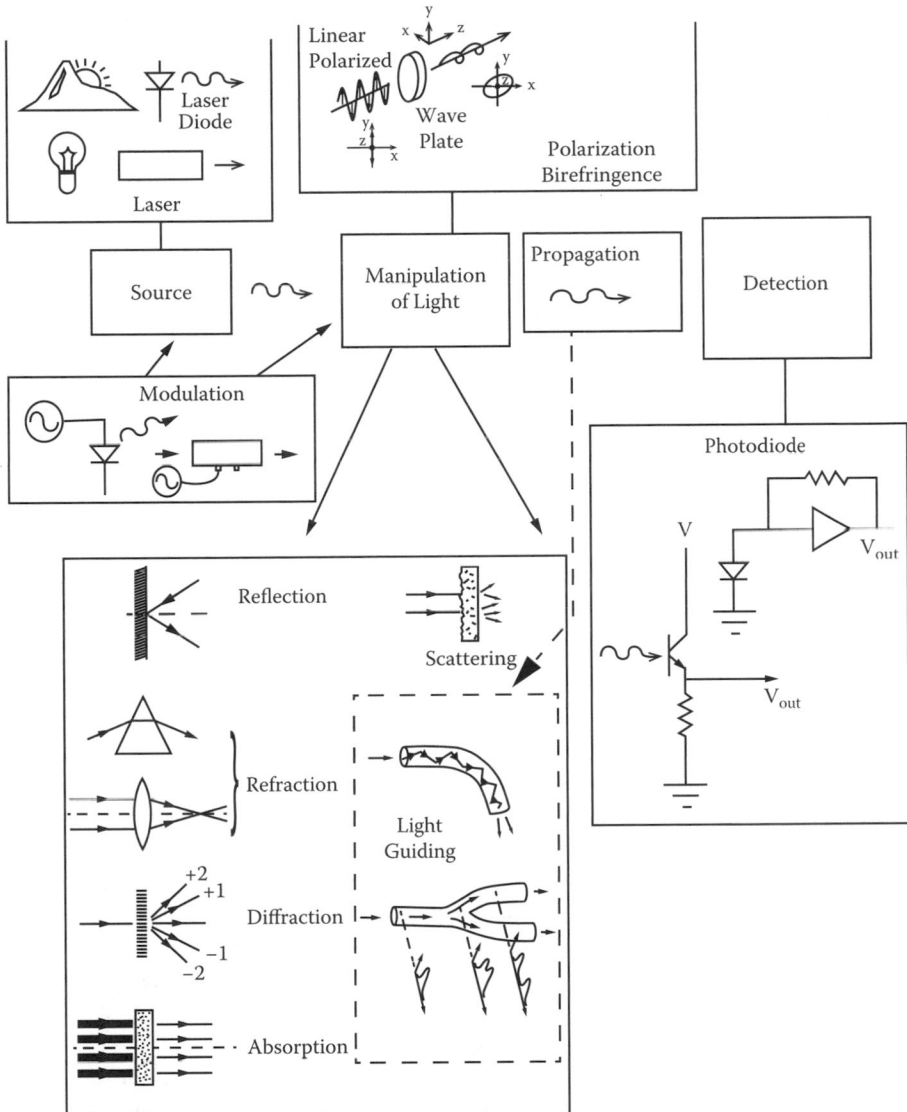

**FIGURE 1.1**   A schematic diagram of some of the components and parts that form an optical system.

are utilized. In fact, depending on the manufacturer, light sources can be specified in a variety of units. For these reasons, conversion from photometric to radiometric units and from radiometric to photometric units is discussed in this chapter. These exercises make it easier to compare light sources from various vendors. For more details in light sources, a textbook in radiometry or photometry is a good resource [1].

In Chapter 3 light detection is discussed, including various detector types, such as photon detectors and thermal detectors, and other topics relating to light detection, such as detector noise and detector figures of merit. The material covered is intended to help the reader understand various options available for light detection and what to look for when choosing a detector. For more detailed understanding of light detection subject, the reader should look into an optics specific text [2–4], other specialized radiometry books [1], or relevant optics handbooks [5].

Chapter 4 discusses manipulation of light. The phenomenon described in this chapter occurs mostly because light encounters a medium or obstruction (with the exception of diffraction, which also occurs as light propagates in free space). The chapter covers reflection, refraction, diffraction and interference, absorption, and scattering. The chapter also contains suggested experiments. The aim of these experiments is to help the reader understand and gain hands-on experience with the topics discussed. The experiments are intended to be performed by materials that are readily available to everyone without the need for any special apparatus. In addition, some of the MATLAB® simulations, such as refraction and diffraction, are included for download and will enable the reader better understand these phenomena.

In Chapter 5 polarization related topics are discussed. Polarized light is encountered both in nature (e.g., scattering from sky), and it is extensively used in optical systems. Polarization phenomenon is used to minimize reflection, and in many instruments such as liquid crystal displays. The chapter describes methods of calculating polarization angles and conditions, and one of the MATLAB simulations also illustrates the effects of polarizers and wave plates to manipulate light polarization. In addition, several suggested experiments are included that will help the reader better understand polarization and the use of polarization components. Further reading on polarization can be found in various optics and physics text, as well as in dedicated books on polarization [6].

Chapter 6 discusses basic principles of geometrical optics, covering ray tracing and formulation based on the assumption that light comprises of optical "rays." Geometrical optics plays an important part in designing optical systems, and understanding how light travels through optical components and systems. In geometric optics, dimensions are much larger than wavelength of light, and light can be estimated in a series of rays. Ray paths are calculated to determine parameters such as image position and aberrations in an optical system. Several methods of ray tracing and primary aberrations of optical systems are discussed. One of the MATLAB simulations illustrates imaging using a single thin lens. For further reading on the subject of geometrical optics and ray tracing, the reader is directed to optics and lens design texts [2–4,7,8].

Chapter 7 describes imaging systems. Topics that related to imaging systems, such as resolution are described, and some imaging systems such as line scanners and two-dimensional imaging are discussed.

Chapter 8 describes guiding light waves. Light guiding is a method to control direction of light in one or more dimensions. Depending on the dimensions, guiding can be treated by means of geometric optics when the dimensions are much larger than the wavelength of light, or can be treated as optical waves when the dimensions are small and approaching light wavelength (e.g., single mode optical fibers). First, the background on light guiding is introduced, then various aspects of it are discussed, such as fiber optics, coupling between waveguides, and active waveguide components. A suggested experiment is included to help the reader understand light guiding. To learn more about light guiding and fiber optics, further information can be found in optics literature as well as in books that cover these topics [9–11].

Chapter 9 describes various topics related to optics, electronics, software, and applications. The chapter covers combining optical systems with electronics and software. Examples are given on how to separate optics and electronics effects to better understand sources of problems. Toward the end of the chapter applications of optical systems are included.

One key area where optics is utilized extensively is sensing. Chapter 10 describes various optical sensing phenomena and different types of sensors are presented. Free-space optical sensors as well as fiber-based optical sensors are discussed. Toward the end of the chapter a brief section gives suggestions on how to choose sensors. Since optical sensing is such a broad topic, there is no one book that covers all sensing applications. Therefore for further reading the reader should find the appropriate text in the area of interest. Many dedicated books in each variety of optical sensor areas can be found in the available technical literature. For example, Dakin and Culshaw's book [12] is dedicated to optical fiber sensors.

The suggested experiments included in some of the chapters was intended to help the reader understand and visualize optical phenomena by performing experiments using materials that are readily available. However some experiments, although simple, would require some special equipment or materials. Chapter 11 describes experiments that are more advanced in nature and sometimes require additional apparatus. However, many of these materials or apparatus (such as electric motors or oscilloscopes) are common in many laboratories. Therefore the reader is encouraged to read this chapter, and if materials required for these experiments are available, then perform the test. If this is not possible, at least reading about these experiments and visualizing them will also be beneficial to further understand optical phenomena.

The optical field is very broad, and many of the topics were either not covered or only briefly glanced. Topics that were not covered and may be too advanced for this book are briefly described in Chapter 12.

## REFERENCES

1. Boyd, R. W. 1983. *Radiometry and the Detection of Optical Radiation*. New York: John Wiley & Sons.
2. Born, M., and E. Wolf. 1999. *Principles of Optics: Electromagnetic Theory of Propagation, Interference and Diffraction of Light*. 7th ed. New York: Pergamon Press.
3. Hecht, E. 2001. *Optics*. 4th ed. Massachusetts: Addison-Wesley.
4. Klein, M. V., and T. E. Furtak. 1986. *Optics*. 2nd ed. New York: John Wiley & Sons.
5. Wolfe, W. L., and G. J. Zissis. 1985. *The Infrared Handbook*. Revised ed. Virginia: General Dynamics.
6. Kliger, D. S., J. W. Lewis, and C. E. Randall. 1990. *Polarized Light in Optics and Spectroscopy*. Boston: Harcourt Brace Jovanovich.
7. Jenkins, F. A., and H. E. White. 1976. *Fundamentals of Optics*. 4th ed. New York: McGraw-Hill.
8. Kingslake, R., and R. B. Johnson. 2009. *Lens Design Fundamentals*. 2nd ed. Orlando, FL: Academic Press.

9. Lee, D. L. 1986. *Electromagnetic Principles of Integrated Optics.* New York: John Wiley & Sons.
10. Okoshi, T. 1982. *Optical Fibers.* New York: Harcourt Brace Jovanovich.
11. Kapany, N. S., and J. J. Burke. 1972. *Optical Waveguides.* New York: Academic Press.
12. Dakin, J., and B. Culshaw. 1988. *Optical Fiber Sensors: Principles and Components.* Vol. 1. Boston: Artech House.

# 2 Light Sources

Both natural and artificial light are used for optics. For example, for outdoor photography, natural lighting is often the choice. Other applications require artificial lighting. Artificial lighting sources that are often used in optical apparatus are

- Incandescent or halogen light—Applications include room lighting, photography, microscopy, and spectroscopy.
- Light emitting diodes—Applications include efficient room lighting, photography, computer screens, microscopy, and spectroscopy.
- Laser diodes—Applications include communication, laser pointers, sensing, laser printers, and CD and DVD players. High-power lasers are used as a pump source for a pulsed power lasers and for optical fiber amplifiers.
- Gas lasers—Applications include holography, stable light source science laboratories, sensing, and medical lasers. High-power gas lasers are used for laser etching and industrial applications.

## 2.1 LASERS (LIGHT AMPLIFICATION BY STIMULATED EMISSION OF RADIATION)

Figure 2.1 shows the basic structure of a laser. In a lasing medium when a photon encounters an atom at an excited state, it stimulates the emission of another photon [1,2]. This phenomenon is repeated as photons travel through the medium and are reflected from the mirrors at the two ends, as shown is Figure 2.1. Once in a while, a photon will escape from the partially reflective mirror, which is the output of the laser. In order for lasing to occur, it requires an energy source, a gain medium (e.g., argon ion gas, a ruby crystal), and feedback (such as reflective mirrors) to allow photons to travel long paths, while stimulating emission of other photons. Examples of energy sources for lasers are electrical energy as a high voltage source for a helium neon (HeNe) laser (similar to a neon lamp), light energy as a flash-lamp used for pumping a Nd-YAG laser with light, and a chemical reaction used in high-energy chemical lasers. Some characteristics of laser light are narrow spectral line width (typically subnanometers), longer coherence length than ordinary light sources, and therefore are suitable for performing interference experiments. Another advantage is that extremely narrow, low-divergence beams of light can be generated. Laser light can be focused to very small spots. Pulsed lasers can generate high-peak powers with extremely short pulses. Lasers currently available include gas lasers for very narrow line width and long coherent length, or high-power [3] and short pulses [4] (which are used for industrial and scientific applications); solid-state lasers that are compact, rugged, and capable of high-speed modulation [5]; and fiber lasers that are

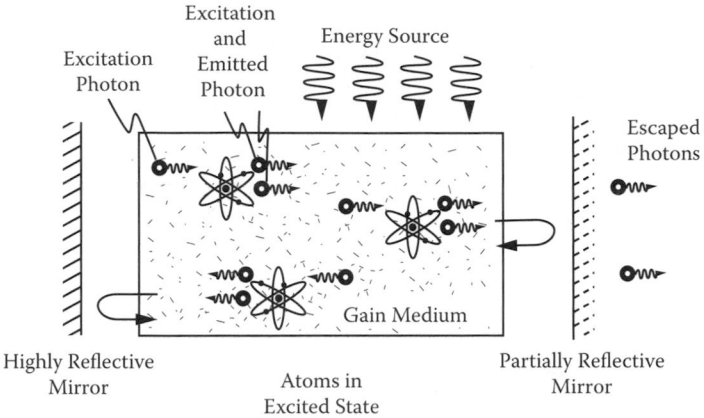

**FIGURE 2.1**   Simplified principle of operation of a laser.

capable of producing ultrashort pulses and high-peak power in a rugged, stable, and compact format.

## 2.2   LIGHT-EMITTING DIODES

Light-emitting diodes (LEDs) are semiconductor devices that emit light at a selective wavelength band. The typical spectral band of single color LEDs is 15 to 50 nm. Other models emit wider bandwidth resembling a white light spectrum. LEDs are a long-life and efficient light source that are widely used for display and lighting applications. Because of the wider spectral bandwidth compared to lasers, their coherence length is much shorter. They have limited applications where light interference is required. However, their use is increasing for several interferometric applications, such as in white light interferometers (instruments used for precise surface topography measurements).

## 2.3   SELECTING LIGHT SOURCES

Choosing optical components, such as light sources, can be challenging, particularly for those who are just getting familiarized with the field of optics. For example, an electronics engineer who is given the task of choosing a light source for a particular application may find him- or herself in a position to select a light source, such as an LED, from a large number of available components. Despite the best efforts by manufacturers to simplify component selection, very often comparisons are difficult between different vendors because of variations in specifications.

To select components from various vendors, one option is to come up with a comparison chart based on the available data from spec sheets, and converting them to similar units.

Another approach is to do a quick search and comparison based few key values, such as wavelength or output color, and maximum current. Next, samples can be requested (if available) from the vendor, test them, then create a comparison chart. For example, if searching for a white LED with a certain size, footprint, and shape, a good starting point is to search for all the components that meet that general requirement. If, for example, optical output is given in different units, a good indicator is current handling capability of the part (generally given in maximum current). Typically, most of the current components that use similar technology and are in the same price range have similar optical output versus current characteristics, and therefore the *maximum current* is a good starting point for a rough comparison. Once a few parts are selected, then a detailed comparison can be performed either by converting the optical outputs to similar units (e.g., lumens [lm] or watts [W]; detailed in upcoming sections) or by requesting samples from the vendors and doing a comparison based on actual measurements (e.g., comparing output optical power using a optical power meter). As a precaution, when measuring components with different formats, characteristics of the component that effect the measurement should be carefully considered. For example, when comparing total optical power from a lensed and nonlensed LED, it is important to have a measurement setup that allows total light collection, because light divergence angle of the two LEDs will be different because the presence or absence of a lens.

For components that are often used for lighting, it is common to find them in photometric quantities, such as in lumens (lm) or lux. However, similar parts may also be characterized in radiometric units, such as in watts (W) or watts per square meter (W/m$^2$).

Often optical output units are given in watts per steradian (W/Sr) or lumens per steradian (lm/Sr), which is dependent on the output angle of the LED, its spectral characteristic, and will be different between a lensed and nonlensed LED. Other times the output is given in milliwatts per square centimeter (mW/cm$^2$). This number depends on the distance where the measurement was made, and again should take into account the size of the spot where it was measured. This information may be available somewhere in the spec sheet. Again, following the aforementioned comparison can minimize the confusion and help determine which component is the best candidate for your design needs.

For a complete optical design, both the characteristics of the light source, the detector, and the medium and components that light encounters before reaching the detector need to be taken into account. However, such preliminary analysis can be a starting point for a basic design that can be tested in the laboratory.

## 2.4 CONVERSION FROM RADIOMETRIC TO PHOTOMETRIC QUANTITIES

Radiometric quantities are physical quantities, whereas photometric quantities reference to the spectral response of the human eye. A comparison of radiometric and photometric quantities is given in Table 2.1.

To convert from a radiometric to a photometric value if the source irradiance is given, use the following relation (adapted from Boyd [6]):

$$PhQ = 683 \int RQ(\lambda) \cdot V(\lambda) \cdot d\lambda \tag{2.1}$$

where *PhQ* is the photometric quantity, *RQ* is the radiometric quantity, and $d\lambda$ is the wavelength interval, and $V(\lambda)$ photopic luminous efficiency, namely, the daytime

**TABLE 2.1**
**Radiometric and Photometric Quantities**

| Radiometric Quantity | Radiometric Unit | Photometric Quantity | Photometric Unit |
| --- | --- | --- | --- |
| Radiant energy | joules (J) | Luminous energy | lumen.second (lm.s); talbot = lm s |
| Radiant energy, density | J/m³ | Luminous energy density | lm s/m³ |
| Radiant power, radiant flux | watt (W) | Luminous flux | lumen (lm) |
| Irradiance, radiant exitance, radiant flux density | W/m² | Illuminance, luminous exitance | lm/m² = lux; (lx) |
| Radiant Intensity | W/sr | Luminous intensity | lumens/steradian; lm/sr; candela (cd) |
| Radiance | W/(m² sr) | Luminance | lm/(m² sr); cd/m² |

adapted spectral response of the human eye [6,7], as shown in Figure 2.2 and given in Table 2.2. The digitized form of Equation (2.1) is given by

$$PhQ = 683 \sum_{n=380nm}^{770nm} RQ_{\lambda n} V_{\Delta\lambda}(\lambda_n)\Delta\lambda \tag{2.2}$$

Note that $V\Delta_\lambda(\lambda_n)$ is digitized $V(\lambda)$ at $\Delta\lambda$ intervals, and is zero outside the range of 380 nm to 770 nm. Therefore there is no need to integrate outside this range.

The value 683 is a conversion factor between lumens and watts [6,7]. (The value 683 comes from the adapted definition of lumen, which is the luminous flux of monochromatic radiation at 555 nm, namely the peak sensitivity wavelength of light-adapted eye. The spectral luminous efficacy is $K_\lambda = K_m V(\lambda)$, where $K_m = 683$ lm/W).

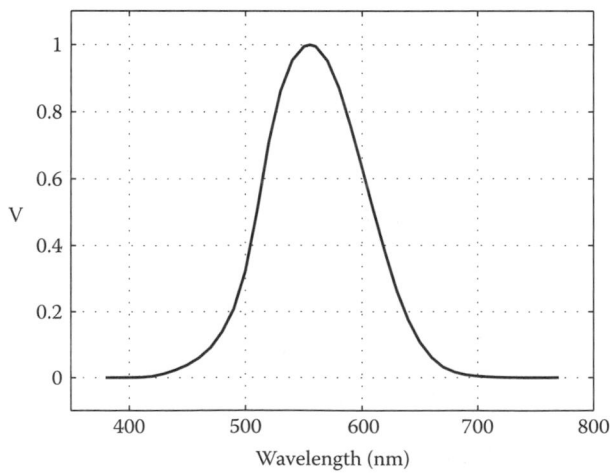

**FIGURE 2.2**  Photopic luminous efficiency, namely, the daytime adapted spectral response of the human eye.

## TABLE 2.2
## Photopic Luminous Efficiency Given at Increments of $\Delta\lambda$ = 20 nm

| Wavelength (nm) | Photopic Luminous Efficiency ($V_\lambda$) | Wavelength (nm) | Photopic Luminous Efficiency ($V_\lambda$) |
|---|---|---|---|
| 380 | 0.00004 | 580 | 0.87 |
| 400 | 0.0004 | 600 | 0.631 |
| 420 | 0.004 | 620 | 0.381 |
| 440 | 0.023 | 640 | 0.175 |
| 460 | 0.06 | 660 | 0.061 |
| 480 | 0.139 | 680 | 0.017 |
| 500 | 0.323 | 700 | 0.0041 |
| 520 | 0.71 | 720 | 0.00105 |
| 540 | 0.954 | 740 | 0.00025 |
| 560 | 0.995 | 760 | 0.00006 |

To convert from radiometric quantity such as watts·m$^{-2}$·nm$^{-1}$, to photometric quantity, such as lumens.m$^{-2}$, use Equation (2.2). An example of such calculation is shown in Table 2.3.

If the units are in milliwatts per square centimeter (mW/cm$^2$), then the corresponding conversion will be in millilumens per square centimeter (mlm/cm$^2$). If the units are in milliwatts (mW), then the corresponding conversion will be in millesimal (mlm).

## TABLE 2.3
## Example of Converting from Radiometric Quantity to Photometric Quantity Given Source Irradiance

| Wavelength (nm) | Photopic Luminous Efficiency ($V_\lambda$), $\Delta\lambda$ = 20 nm | Example Source Irradiance [mW/(m$^2$. nm)] | Irradiance × V × $\Delta\lambda$ |
|---|---|---|---|
| 460 | 0.06 | 0 | 0 |
| 480 | 0.139 | 0.12 | 0.3336 |
| 500 | 0.323 | 0.5 | 3.23 |
| 520 | 0.71 | 0.8 | 11.36 |
| 540 | 0.954 | 0.6 | 11.448 |
| 560 | 0.995 | 0.05 | 0.995 |
| 580 | 0.87 | 0 | 0 |

Sum of Irradiance = 2.07 mW/m$^2$

| | |
|---|---|
| Sum(Irradiance × $V$ × $\Delta\lambda$) | 27.3666 |
| Photometric Quantity: | 18693 mlm/m$^2$ |
| 683 × Sum(Irradiance × $V$ × $\Delta\lambda$) | |
| | 18.7 lm/m$^2$ |

## 2.5 CONVERSION FROM PHOTOMETRIC TO RADIOMETRIC QUANTITIES

To convert from photometric quantities (e.g., lumens) to radiometric quantities (e.g., watts) for a source that falls within the visible part of the spectrum, the reverse process is used from the one described in radiometric to photometric conversion. The information needed to do this is a relative or normalized spectral response and the photometric quantity. Therefore the radiometric quantity (RQ) can be estimated using the following equation:

$$RQ = \frac{PhQ \cdot \sum_{n=380nm}^{770nm} I_R(\lambda_n)}{\Delta\lambda \cdot 683 \sum_{n=380nm}^{770nm} I_R(\lambda_n) \cdot V_{\Delta\lambda}(\lambda_n)} \tag{2.3}$$

where $I_R(\lambda)$ is the relative spectral output of the source. Equation (2.3) represents the discrete version of calculation, and $I_R(\lambda_n)$ and $V_{\Delta\lambda}(\lambda_n)$ are at dl increments (e.g., at 10 nm increments).

Note that Equation (2.3) is only valid for a source that falls completely within the range $V(\lambda)$ covers, namely, between 380 nm and 770 nm. For example, if a source emits in the visible as well as in the ultraviolet (UV), then the total power calculated using Equation (2.3) will not reveal the UV emission power, and therefore Equation (2.3) is not valid.

## 2.6 THERMAL SOURCES

Thermal sources are light sources that emit in the visible and infrared wavelengths. Emission wavelength is dependent on the temperature of the object. For example, a hot incandescent light emits light in the visible and near infrared, whereas a warm body emits thermal radiation around 9 microns (see Figure 2.3c). Because of this characteristic, thermal detectors (discussed in Chapter 3) are used for night vision in the absence of natural lighting. Many thermal source emissions can be estimated by a blackbody radiation curve.

## 2.7 BLACKBODY RADIATION

A hot object emits electromagnetic radiation that has a characteristic emission spectra that is related to the temperature of the object. A true blackbody emission can be obtained from an enclosure (such as a hot metal hollow box) with a small hole, where the emission spectra at the output is given by [6]

$$I(\lambda, T) = \frac{2hc^2}{\lambda^5} \frac{1}{e^{\frac{hc}{\lambda kT}} - 1} \tag{2.4}$$

where $h$ is the Planck constant ($h = 6.63 \times 10^{-34}$ J.s), $c$ is the speed of light in vacuum ($c = 3 \times 10^8$ m/s), and $k$ is the Boltzmann constant ($k = 1.38 \times 10^{-23}$ J/K). Output of many thermal sources can be estimated using Equation (2.4).

Figure 2.3 shows emission spectra at three color spectra: 2800 K (household incandescent light bulb), 5800 K (temperature of the surface of the sun), and 37°C (average human body temperature).

Figure 2.3 plots are in normalized units. The factor ($2hc^2/\lambda^5$) in Equation (2.4) determines the amplitude. As expected, incandescent light bulb and sun spectral emission (Figure 2.3a,b) cover the visible range of the spectrum (400 to 700 nm). However, emission from a warm object at 37°C emit in the infrared spectrum, peaking around

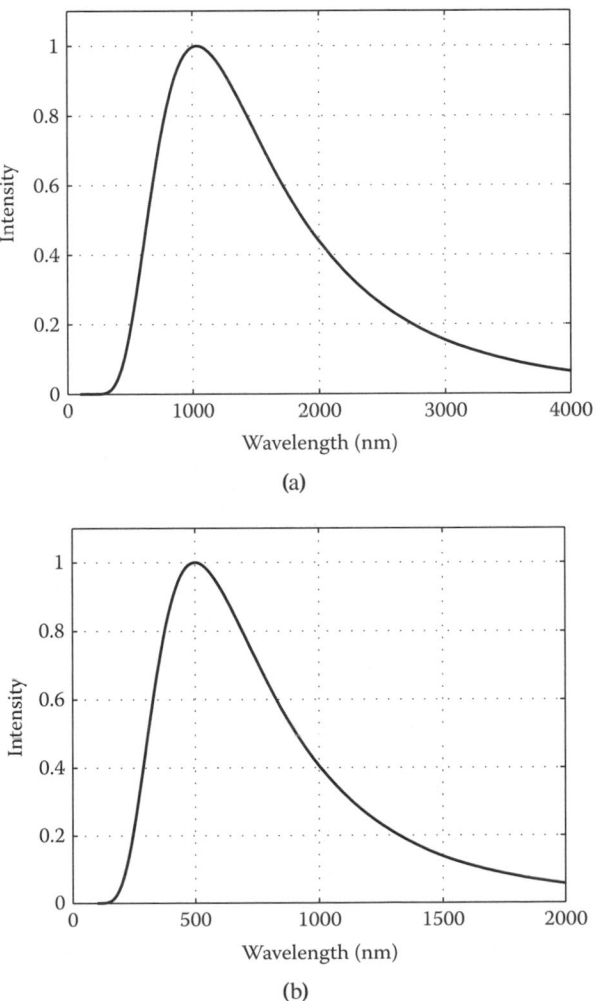

(a)

(b)

**FIGURE 2.3**  Normalized intensity of blackbody emission spectra at three different temperatures. Emission at: a) 2800 °K, b) 5800 °K, and c) 310 °K, or 37 °C. *(continued)*

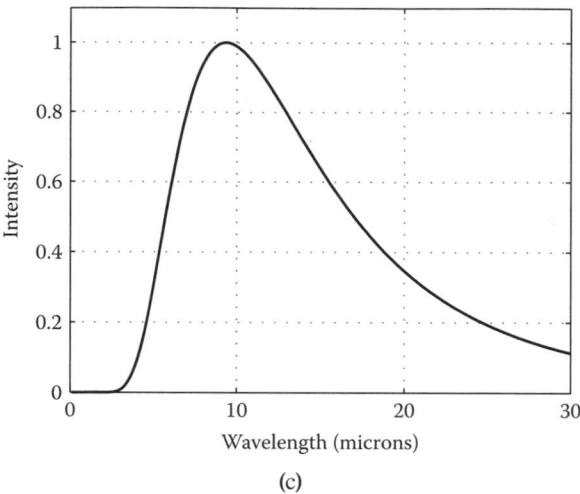

(c)

**FIGURE 2.3**    *(continued)* Normalized intensity of blackbody emission spectra at three different temperatures. Emission at: a) 2800 °K, b) 5800 °K, and c) 310 °K, or 37 °C.

9 microns. Because of this, thermal imagers can image humans even in complete darkness.

## REFERENCES

1. Yariv, A. 1991. *Optical Electronics.* 4th ed. Philadelphia: Saunders College Publishing.
2. Verdeyen, J. T. 1995. *Laser Electronics.* 3rd ed. Englewood Cliffs, NJ: Prentice Hall.
3  Injeyan, H., and G. Goodno. 2011. *High Power Laser Handbook.* New York: McGraw-Hill Professional.
4. Binh, L. N., and N. Q. Ngo. 2010. *Ultra-Fast Fiber Lasers: Principles and Applications with MATLAB® Models (Optics and Photonics).* Boca Raton, FL: CRC Press.
5. Coldren, L. A., Scott W. Corzine, and M. L. Mashanovitch. 2012. *Diode Lasers and Photonic Integrated Circuits.* 2nd ed. New York: Wiley.
6. Boyd, R. W. 1983. *Radiometry and the Detection of Optical Radiation.* New York: John Wiley & Sons.
7. Mahajan, V. N. 1998. *Optical Imaging and Aberrations, Part I: Ray Geometrical Optics.* Washington: SPIE Press.

# 3 Light Detection

An important part of an optical system is to convert light to a signal that then can be analyzed or stored. Examples of light detectors are

- Cone and rod cells in the retina in the human eye
- Optical recording materials, such as silver halide film used in analog photography
- Photodiodes used as optical receivers in fiber-optic communication systems
- A linear detector array used in a fax machine or a paper scanner
- A two-dimensional detector array used in a digital camera

The basic principles of using a photon detector involve detection of light and converting the light intensity signal to current or voltage via an amplifier.

In a detector array a number of methods are used to convert the light intensity distribution to an electrical signal to be processed by electronics. Some examples of array detectors are:

- A CMOS (complementary metal-oxide semiconductor) detector array that includes the switching electronics, where the light intensity distribution is converted to an array of electrical pulses, each pulse corresponding to light intensity at a specific pixel on the detector array
- An infrared (IR) detector array, such as an HgCdTe [1] detector, connected to a read-out integrated electronic circuit (ROIC) via bump-bonding

## 3.1 PHOTON DETECTORS

Photon detectors directly respond to incident photons. The absorbed photons change the electronic characteristic of the detector [2]. The effect could be emission of electrons from a metal surface, or semiconductor detectors, where electron–hole pairs are generated when photons are absorbed.

Photon detectors are generally much more sensitive than thermal detectors; thermal detectors, however, often cover broader spectral bands than photon detectors.

Photodiodes, phototransistors, and photomultipliers are examples of photon detectors. In a photodiode, the incident photons generate a photocurrent proportional to the light intensity. Photodiodes (PDs) can be used either with a reverse bias (photoconductive mode) or unbiased (photovoltaic mode). An example of a photodiode connected to a transimpedance amplifier (TIA) circuit is shown in Figure 3.1, where the output of the amplifier circuit is the product of the photocurrent ($I_{PD}$) and the feedback resistor ($R_f$):

$$V_{out} = -I_{PD}R_f \qquad (3.1)$$

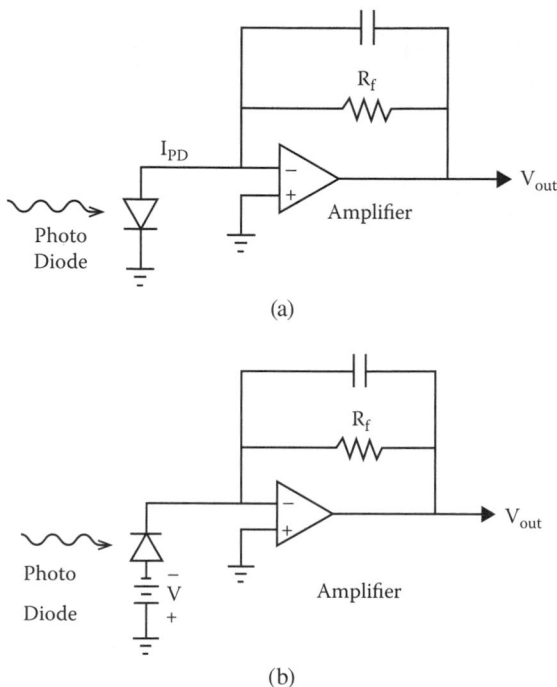

(a)

(b)

**FIGURE 3.1**   (a) Photodiode in a transimpedance amplifier circuit in a photovoltaic mode $V_{out} = -I_{PD}R_f$. (b) Photodiode amplifier in a photoconductive mode.

The TIA converts the current produced by the PD to a voltage. There are two different modes of operation of photodiod TIA: photovoltaic and photoconductive [2]. In a photovoltaic mode (Figure 3.1a), no voltage is applied to the PD. In a photoconductive mode (Figure 3.1b), the PD is reverse biased. The purpose of reverse biasing is to increase the signal response, but the drawback is a higher noise. The voltage output of the TIA is also given by

$$V_{out} = \eta P R_f \tag{3.2}$$

where $P$ is the optical power (in watts), and $R_f$ is the value of the feedback resistor, and $\eta$ is the detector sensitivity (in amperes per watt [A/W]). The higher the value of $R_f$, the higher is the voltage output. However the noise increases as well but at a slower rate. Namely, noise increases by $\sqrt{R_f}$. In addition, the higher the $R_f$, the lower the bandwidth (slower frequency response). For an Si detector, $\eta$ increases from 350 nm and peaks at around 1100 nm, and then dramatically drops. A typical number is 0.4 A/W at 633 nm (photodiode response curves can be found in the vendor specifications). The feedback capacitor is used to prevent gain peaking, but high capacitance can also limit the frequency response.

Another example of a photon detector is a phototransistor. Phototransistors produce large enough signals using only a resistor circuit, as shown in Figure 3.2.

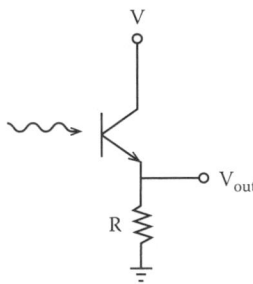

**FIGURE 3.2**   Phototransistor in a circuit.

Phototransistors have slower response and less sensitivity than photodiodes, but require no amplifier circuits. Furthermore, unlike a phototransistor, when a large gain is needed in a photodiode circuit, proper shielding and grounding is required to minimize noise.

## 3.2   THERMAL DETECTORS

Thermal detection is achieved when incident optical radiation is absorbed, causing a rise in the temperature of the detector element, which is then converted to an electrical signal. For example, a bolometer [1,2] is a device where the incident light is absorbed, causing a rise in the temperature, which in tern causes change in resistance. The change in resistance is detected with an electrical circuit, resulting in a signal that is proportional to thermal radiation. An extension of a bolometer is a microbolometer, where an array of microfabricated bolometers is used to image the incident thermal radiation. Many modern thermal cameras use microbolometer technology.

Another type of thermal detector is called a Golay cell [2], which consists of an enclosure (cell) filled with gas and an absorbing target inside the cell. One wall of the cell is made of a flexible membrane. Light absorbed by the target causes temperature change and the gas expands. This causes the membrane to flex. The change in membrane shape is monitored by light reflection.

## 3.3   NOISE IN PHOTODIODES

Noise in photodiodes is the sum of Johnson noise [3], noise due to dark current, and noise due to light current. The total noise current is given by [4]

$$i_n = \sqrt{i_J^2 + i_{Dark}^2 + i_{Light}^2} \tag{3.3}$$

The first term is Johnson noise, which is due to thermal agitation of electrons in a resistor that results in a noise current [2] given by

$$i_J = \sqrt{\frac{4k \cdot T \cdot \Delta f}{R_{Sh}}} \tag{3.4}$$

where $k$ is the Boltzmann's constant, $T$ is the absolute temperature of the detector element, and $\Delta f$ is the noise bandwidth, and $R_{Sh}$ is the shunt resistance.

The other term in Equation (3.3) is given by

$$i_{Dark} = \sqrt{2 \cdot q \cdot I_{Dark} \cdot \Delta f} \qquad (3.5)$$

where $q$ is the electron charge and $I_{Dark}$ is the dark current. The third term in Equation (3.3) is given by

$$i_{Light} = \sqrt{4 \cdot q \cdot I_{Light} \cdot \Delta f} \qquad (3.6)$$

## 3.4   PHOTODETECTORS FOR LOW-LIGHT LEVEL DETECTION

When attempting to detect extremely low levels of light, standard photodiodes may not be sufficient, because to obtain an electrical signal of a measurable level, the signal has to be amplified. Attempting to amplify extremely low signals will also amplify noise. To overcome this issue, there are some optical detectors that are made to produce large signals from extremely low light levels.

A photomultiplier tube (PMT) is a photodetector that can detect extremely low levels of light. It has several stages of amplification, where a photo-generated electron is multiplied. Because PMTs are extremely sensitive, they are to be handled with extra caution, being careful not to damage them by exposure to bright light when powered on. Examples of uses of PMTs are for low-light level florescent measurement and single photon detection.

An avalanche photodiode is another very sensitive photodetector providing large signals from extremely low light levels.

## 3.5   INTEGRATING SPHERES FOR LIGHT MEASUREMENT

To measure total optical power from a source that has a wide range of angles, such as emission from an LED, an apparatus called an integrating sphere can be used to collect the total light emission. A basic diagram of on integrating sphere is shown in Figure 3.3. Measurements made with an integrating sphere are insensitive to detector position, and is therefore suitable for nonuniform and diverging source measurements. By the time light is received at the photodetector, it is scattered from the inside walls of the integrating sphere and is hightly attenuated, which is a useful feature for making high-power laser measurements.

One of the openings of the integrating sphere is used as a light input port. The other opening is for mounting the photodetector. The sphere is coated with a material or paint with a broad spectral reflectivity (such as a white diffuse

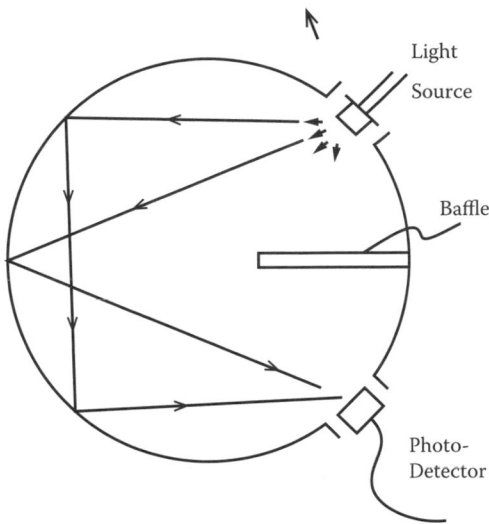

**FIGURE 3.3**  Schematic diagram of an integrating sphere.

coating), and generally has a uniform reflection coefficient in the spectral range of interest. In some instances, it is desirable to detect only scattered light from the walls of the integrating sphere and avoid direct light emitting from the source to reach the detector. In this case the light input and photodetector ports are selected such that there is a baffle in between the ports to block direct light from source to detector.

## 3.6  LOCK-IN AMPLIFICATION FOR DETECTING LOW-LIGHT LEVEL SIGNALS

One method of detecting a signal that is embedded in noise is to use a lock-in amplifier. An example is to detect an optical signal in the presence of high ambient light. A lock-in amplifier is an instrument that synchronizes a detected signal (e.g., a signal from a photodiode) with a modulated light source. The modulation can be achieved either using a mechanical chopper or by modulating the light source (see Figure 3.4). A good source for further understanding of the lock-in amplification process can be found in the manufacturer's technical notes (see, for example, Stanford Research Instruments website for tutorials on lock-in amplifiers, www.thinksrs. com/downloads/PDFs/Application Notes/About LIAs. pdf). Many of the old lock-in amplifiers consisted of analog electronics. With the advent of digital electronics, current lock-in amplifiers are digital, using digital signal processing as a means of synchronizing the modulated source with the detected signal. Using lock-in amplifiers enables detection of signals with signal level below signal-to-noise ratio (SNR) of less than 1.

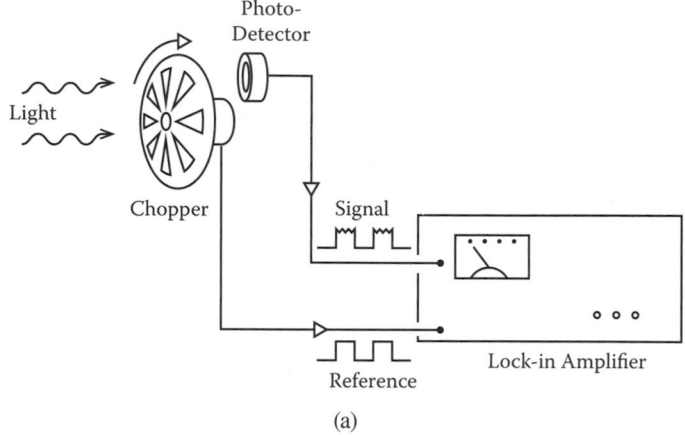

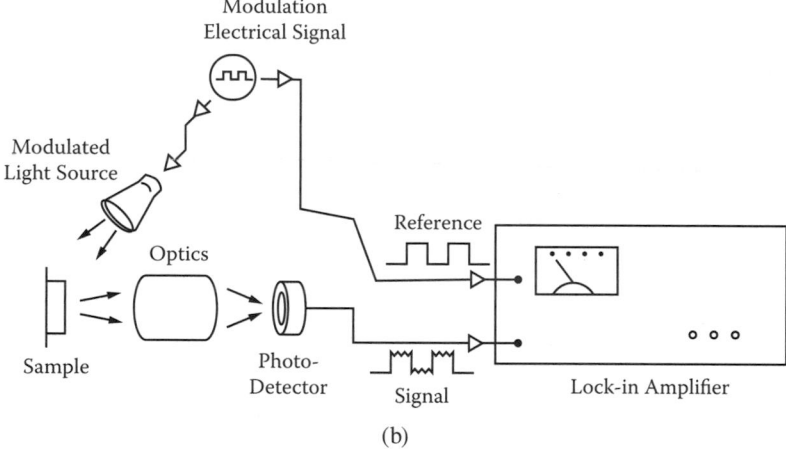

**FIGURE 3.4**   Lock-in detection scheme using (a) a mechanical chopper and (b) by modulating the light source.

## 3.7   DETECTOR FIGURES OF MERIT

Some of the figures of merit of detectors are detectivity ($D^*$, pronounced D-star). For thermal detectors, another figure of merit is the noise equivalent thermal temperature difference (*NEdT*).

### 3.7.1   DETECTIVITY

Detectivity is used because it takes into account detector size, noise at a given bandwidth, and detector area, and therefore a variety of detector technologies can be compared with this figure of merit. Normalized detectivity, $D^*$, is given by

$$D^* = [(A_D \times \Delta f)^{1/2}]/NEP \tag{3.8}$$

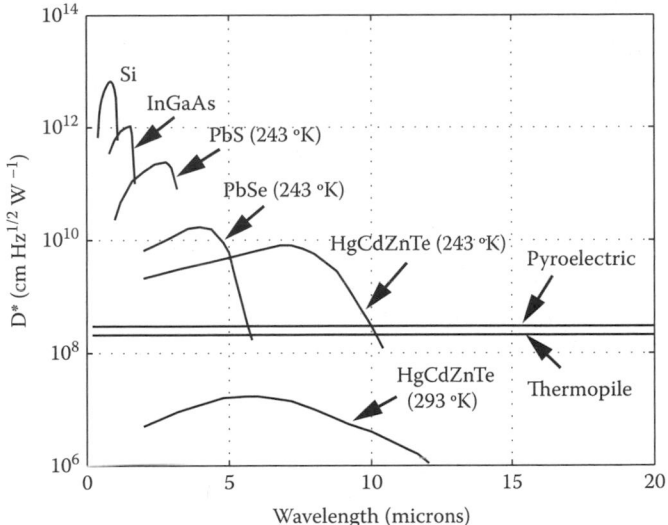

**FIGURE 3.5**  Detectivity (D*) of various optical detector technologies. (Data from Wolfe, W. L., and G. J. Zissis, 1985, *The Infrared Handbook*, revised ed., Virginia: General Dynamics; and Boyd, R. W., 1983, *Radiometry and the Detection of Optical Radiation*, New York: John Wiley & Sons.)

where $A_D$ is the detector area, $\Delta f$ is the detection bandwidth, and *NEP* is the noise equivalent power, which is the incident optical power to produce signal-to-noise ratio equal to unity, (SNR = 1), and is given by [2]

$$NEP = V_n/R \tag{3.9}$$

where $V_n$ is the noise voltage and $R$ is the responsivity of the detector. Example of D* plots for variety of detector technologies are shown in Figure 3.5.

### 3.7.2  NOISE EQUIVALENT TEMPERATURE DIFFERENCE

A key parameter that indicates effectiveness of a thermal sensor is the noise equivalent temperature difference (*NEdT*), which is defined as

$$NEdT = V_n \frac{\Delta T}{\Delta V_s} \tag{3.9}$$

where $V_n$ is the RMS noise voltage, $\Delta T$ is the temperature difference and $V_s$ is the signal due to change in temperature. $\Delta T$ is typically measured by taking measurements at two different temperatures. When *NEdT* is given, it is customary to also define *f/#* (see Chapter 7 for definition) of the measurement, because signal amplitude of the measurement depends on the *f/#* of the system.

## 3.8  SPECTROMETERS

A spectrometer is an instrument that measures optical radiation. It outputs a measurement of signal level versus wavelength (or in some instruments, inverse of wavelength). A spectrometer consists of optics to collect light, and one or more diffraction gratings. The grating diffracts light depending on wavelength. An alternative to diffraction grating is to use a refractive prism. In a *refractive* element, such as a prism, *the shorter the light wavelength, the larger is the refracted light angle.* Namely, the refraction angle of blue is larger than red. On the other hand, in a *diffractive* element such as a diffraction grating, *the longer the light wavelength, the larger is the diffracted light angle.* Namely, the diffraction angle of red is larger than blue. Refraction and diffraction are explained in Chapter 4. Earlier spectrometers consisted of a mechanism to rotate the grating and a single element photodetector that detected diffracted light. Therefore to obtain spectral response, the grating had to be rotated and data captured for each position of the grating that corresponded to a particular wavelength. Many modern-day spectrometers utilize linear array detectors instead of a rotating mechanism, and the detector array captures the diffracted light and produces spectral measurement in real-time. Spectrometers are discussed further in Chapter 12.

## REFERENCES

1. Wolfe, W. L., and G. J. Zissis. 1985. *The Infrared Handbook.* Revised ed. Virginia: General Dynamics.
2. Boyd, R. W. 1983. *Radiometry and the Detection of Optical Radiation.* New York: John Wiley & Sons.
3. Horowitz, P., and W. Hill. 1989. *The Art of Electronics.* 2nd ed. Cambridge, UK: Cambridge University Press.
4. Young, P. H. 2003. *Electronic Communication Techniques.* 5th ed. Upper Saddle River, NJ: Prentice Hall.

# $4$ Manipulation of Light

This chapter describes light that is manipulated by a medium or obstruction. Many of the optical phenomena we observe are a direct cause of light encountering an object or a medium other than propagating in vacuum.

## 4.1 REFLECTION

### 4.1.1 REFLECTION FROM FLAT MIRROR

As observed in our daily experience, light rays reflect, namely, change direction and travel back, when they encounter a smooth surface, such as glass or a polished metal. The reflection angle from a polished surface, such as glass or a front-surface mirror, is equal to the incident angle, both measured from the surface normal, as shown in Figure 4.1. Namely,

$$\theta_r = \theta_i \tag{4.1}$$

where $\theta_i$ is the incident angle, and $\theta_r$ is the reflection angle.

For a curved reflective surface, the same law applies, but since the normal to the surface varies across the curved surface, reflection is dependent on the position of the incident light ray to the curved surface. For example, parallel incident light rays reflect to different angle, as shown in Figure 4.2 and Figure 4.3 for a concave and convex mirror, respectively. When parallel light rays (also referred to as collimated light) are incident on a concave surface (Figure 4.2), rays are convergent into a focus. Conversely, a convex mirror causes collimated light to expend, as shown in Figure 4.3. The insets show details of the angle of incidence and reflection with respect to the surface normal.

A spherical concave mirror with radius of curvature $r$, focuses light at a focal distance $f$ from the mirror. The focal length $f = r/2$ is illustrated in Figure 4.4.

## 4.2 REFRACTION

When light rays pass through a dielectric medium, the speed of light is slower in the medium than in air. For a lossless medium, such as a clear glass, the speed of light in the medium is given $v_{in} = c/n$, where $c$ is the speed of light in vacuum ($3 \times 10^8$ m/s), and $n$ is the refractive index. Light rays traveling from one medium to another bend at the interface. This is called refraction, and the angle relation between the incident and refracted rays is given by Snell's law:

$$n_1 \sin \theta_1 = n_2 \sin \theta_2 \tag{4.2}$$

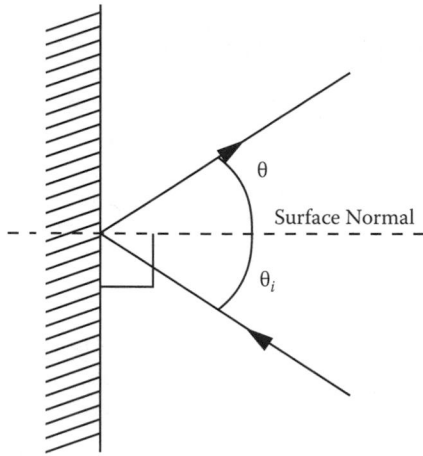

**FIGURE 4.1**  Reflection from mirror.

where $n_1$ and $n_2$ refer to the refractive index in medium 1 and 2, and $\theta_1$ and $\theta_2$ are the angle of incidence and refraction in medium 1 and 2, respectively, as shown in Figure 4.5.

The refractive index $(n)$ is wavelength dependent. For glass in the visible spectrum, the refractive index is larger at shorter wavelengths (e.g., for blue) than at longer wavelengths (e.g., for red). When a monochromatic ray of light travels through a prism, light bends as it travels according to Snell's law (Equation 4.2) at both

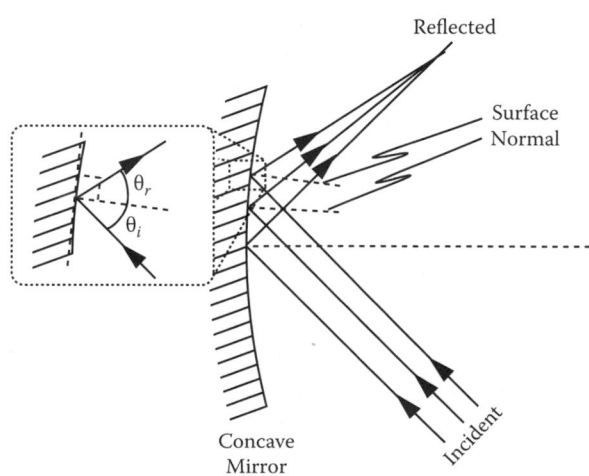

**FIGURE 4.2**  Off-axis reflection from concave mirror.

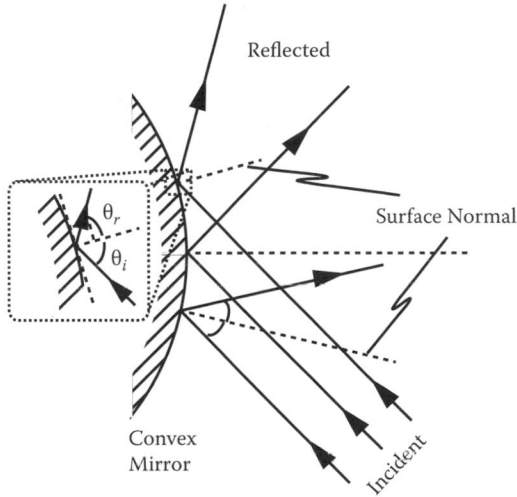

**FIGURE 4.3**   Off-axis reflection from convex mirror.

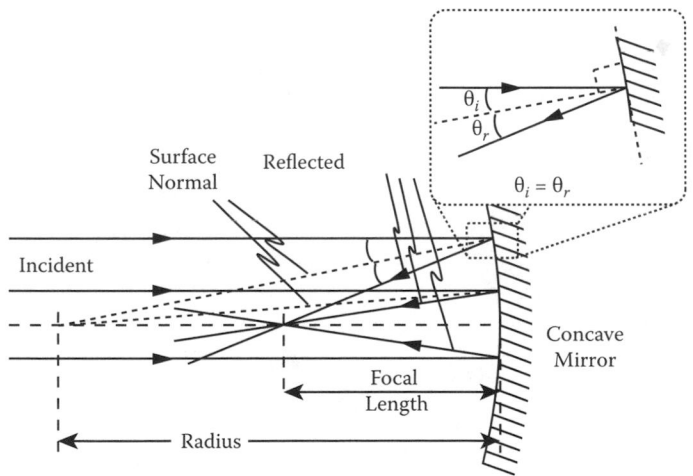

**FIGURE 4.4**   On-axis reflection from convex mirror.

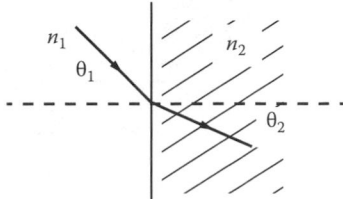

**FIGURE 4.5**   Refraction from low refractive index medium ($n_1$), such as air, to higher refractive index ($n_2$) medium, such as glass.

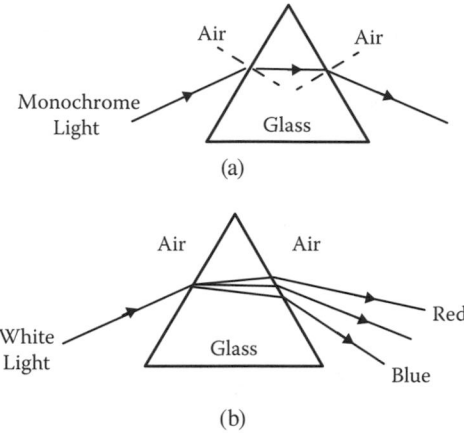

(a)

(b)

**FIGURE 4.6**  Light refraction through a prism (a) from monochromatic light rays and (b) from dispersion from white light.

faces of the prism, as shown in Figure 4.6a. If, instead of a monochromatic light, a ray of white light travels through a prism, because of the dispersion (wavelength dependence of refractive index), white light splits into the rainbow spectrum from red to blue, as shown in Figure 4.6b, because the refractive index for blue is larger than red.

When light rays encounter a curved surface of a material of different refractive index, such as a lens, light also bends according to the Snell's law (Equation 4.2), with the angle of incidence and refraction measured with respect to the normal to the surface (see insets in Figure 4.7). But since the normal to the surface varies across the curved surface, refraction angle is dependent on the position of the incident light ray. This is the reason why positive lenses have the ability to focus light. Figure 4.7 illustrates this fact, where off-axis rays bend toward a spot, whereas on-axis rays that

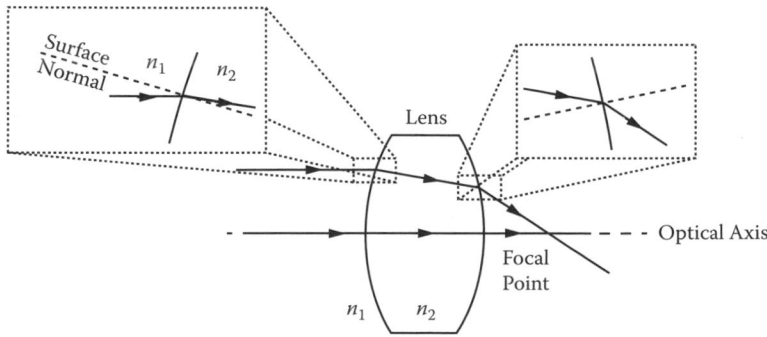

**FIGURE 4.7**   Refraction from thick lens of refractive index n2 (e.g., $n_2 = 1.5$) surrounded by air ($n_1 = 1$). Insets show details of light refraction at the air/glass interface. Dotted lines are the surface normal to the lens surface.

are normal to the surface of the lens go through without bending. Both sets of rays converge to a focal point.

## 4.2.1 REFLECTION COEFFICIENT AT DIELECTRIC INTERFACE

When light propagates between two dielectric interfaces (such as air to glass interface), part of it is reflected. (Light consists of electric and magnetic fields, hence the name electro-magnetic radiation.) The reflection coefficient is the ratio of the amplitude of the reflected and incident electric field. Similarly, the transmission coefficient is the ratio of the amplitude of the reflected and incident electric field. The reflection and transmission coefficients (for normal incidence) are given by

$$r = \frac{n_1 - n_2}{n_1 + n_2} \tag{4.3}$$

$$t = \frac{2n_1}{n_1 + n_2} \tag{4.4}$$

where subscripts 1 and 2 refer the incident and transmitted media. The reflected ($R$) and transmitted power ($T$) is the square Equation 4.3 and Equation 4.4, respectively ($R = |r|^2$, $T = |t|^2$). For example, there is a 4% reflection ($R = 0.04$) at the air ($n = 1.0$) and glass ($n = 1.5$) interface. These coefficients are valid for normal incidence. At nonnormal incidence, reflection and transmission coefficients will depend on incident polarization, and are described in Chapter 5.

## 4.3 DIFFRACTION

One of the properties of light is that it diffracts as it propagates in space. For example, even a "parallel" beam of light will expand due to diffraction as it propagates through space, even if it is unobstructed. Diffraction can also be observed when light encounters an object such as a slit or a sharp edge. The reason for including diffraction as part of this chapter on manipulation of light is because most of the diffraction we will discuss is a result of introducing an object into the light path.

Diffracted light intensity can be calculated in a number of ways. One method is to use a finite element method to observe light behavior at any distance, namely, by solving a wave equation. A simpler method is to use Fresnel or Fraunhofer approximation, which are closed form equations. Very often, the region of interest is in a near field (Fresnel region [1]) or far field (Fraunhofer region [1]), and therefore it is possible to calculate diffraction pattern using close-form equations. These approximations are valid based an object feature, light wavelength, and distance. Appendix B describes Fresnel and Fraunhofer diffraction formulation, and the conditions where these formations are valid. More details on diffraction can be found in Goodman [1].

It should be noted that both Fresnel and Fraunhofer equations have a form of Fourier transform [1]. Therefore making use of fast Fourier transform (FFT) calculation capabilities of computational software, such as MATLAB®, enables visualization of a diffraction pattern.

Diffracted light amplitude in the far field (Fraunhofer) can also be calculated by Fresnel equation, where the oscillating phase factor inside the equation is nearly constant.

To get familiarized with diffraction, use the diffraction simulation listed in Appendix A. These simulations are valid for both Fresnel and Fraunhofer regions. To further understand diffraction and to visualize it, perform the single and double slit diffraction experiments suggested near the end of this chapter.

### 4.3.1 Diffraction Gratings

When light encounters a periodic structure that is on the order of wavelength, it diffracts. This phenomenon is observed when viewing a compact disk (CD). The diffracted light splits white light into different colors of the spectrum. Diffraction from a periodic structure is given by

$$2\Lambda \cdot sin(\theta) = n\lambda \qquad (4.5)$$

where $\Lambda$ is the period of the periodic structure, $\theta$ is the angle of diffraction, l is the optical wavelength, and $n$ is an integer, where $n = \pm1, \pm2, \pm3 \dots$ refer to first, second and third diffraction orders, respectively. As indicated in the above equation (4.5), longer wavelengths diffract at higher angles.

## 4.4 INTERFERENCE

Interference of light is due to the wave nature of light. Much like water waves, interference is observed with light waves. Interference is observed in a region where light rays are coherent. An example of interference observed in nature is the colorful sheen on a soap bubble, where light from inside and outside surfaces of the soap bubble interfere, creating the observed colors. With the advent of lasers and availability of long coherence length light source, interference phenomena have been used in many areas of science in technology. Some of the areas where interference is used are

- Interferometry, such as for metrology and surface topography measurements.
- Holography, where coherent light scattered from an object interferes with a reference light from the same source, and interference fringes are recorded on a photosensitive film. After recording and processing the film, the hologram can be viewed to reveal a three-dimensional object. Unlike photography, where only amplitude is recorded onto the film, holography enables recording of amplitude and phase, representing three-dimensional objects onto a two-dimensional media (phase and amplitude are recorded on interference fringes that are modulated by the phase variation of the light scattered from the object beam).
- Thin film metrology, to determine layers of film.
- Thin film coatings [2]—Due to the interference of light rays from interfaces, it is possible to make antireflection coatings that are found in many of the common objects used daily, such as cameras and eyeglasses.

Experiment 4.2 illustrates generation of interference fringes using laser light (e.g., light from a laser pointer) reflected from both sides of a glass.

## 4.5  ABSORPTION

One of the phenomena that reduces light intensity as it transmits through a medium is absorption. When light is absorbed, its energy is converted into heat motion of the absorbing molecules [3]. This is why a dark glass that is sitting in the sun will get hot much faster than a clear glass. Examples of absorbing media are colored plastic and glass, and dyes used for coloring liquids. Many of the dyes have specific wavelength absorption band. For some applications in optics it is desired to have media that absorb uniformly in a wide range of wavelengths. These components are often referred to as neutral density filters.

As light propagates in absorbing media, its intensity ($I_o$) is reduced to I, namely

$$I = I_o e^{-\alpha l} \tag{4.6}$$

where $l$ is the distance light traveled in the absorbing medium with absorption coefficient of $\alpha$. Light transmission ($T$) through absorbing medium is given by

$$T = \frac{I}{I_o} \tag{4.7}$$

$T$ has a value between 0 and 1 indicating how much light gets transmitted through a medium. One of the measures given for absorption is optical density ($OD$), which is defined as

$$OD = -\log_{10}(T) \tag{4.8}$$

or

$$T = 10^{-OD} \tag{4.9}$$

The larger the $OD$, the more the amount of light is absorbed.

It should be noted that in some earlier text, such as in Jenkins and White [3], optical density is defined as the measure of refractive index. It is more common, particularly with optical component vendors, for optical density to refer to light absorption, as described earlier.

An example of absorption with a colored plastic film is shown in Figure 4.8. Light scattered from red and blue marks on the paper pass through a red plastic filter, and only blue is absorbed, while red is transmitted. The brightness of the red mark is nearly equal to the brightness of the white background when viewed through the red filter, whereas the blue mark looks dark.

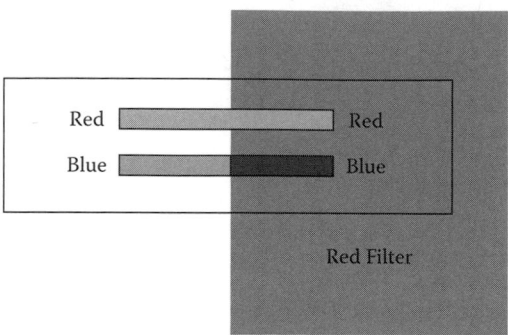

**FIGURE 4.8**    Red and blue print absorbed by a red filter.

## 4.6 DIFFUSERS AND SCATTERING

In a scattering medium, there are randomly oriented variations in the medium, such as absorbers, discontinuities, variations in refractive index, or randomly varying surface height. When light is incident in a random medium, it scatters at random orientations, which is determined by the characteristics of the medium. Examples of a scattering medium are diffusers, light scattering from the atmosphere, and scattering from fluid such as milk. Diffusers can be used to generate uniform illumination lighting.

For detailed description of various types of scattering and their physical origins, the reader is referred to Born et al. [4].

### 4.6.1 SMALL PARTICLE SCATTERING

When light encounters suspended particles that are small compared to the wavelength of the light, then scattering is wavelength dependent. In fact, the scattered intensity is proportional to $1/\lambda^4$, where $\lambda$ is the wavelength of light. This is called Rayleigh scattering, and it is the reason why the sky is blue or why smoke appears bluish when illuminated with white light [3,4]. Because of the $\lambda^{-4}$ dependence of scattering, blue light (short wavelength) is scattered at much more than red light (long wavelength), therefore the sky appears blue during the day. For the same reason, the horizon looks red-orange right after sunset, because blue light gets scattered, and light reaching the observer contains the longer wavelengths of the spectrum, and therefore appears red or orange. In addition to wavelength dependence, scattered light is also polarization dependent, a phenomenon that honeybees use to navigate to their food source [5].

### 4.6.2 LARGE PARTICLE SCATTERING

When particles are larger than wavelength of light, scattering is wavelength independent, and scattered light appears white (such as clouds that contain water molecules that are larger than the wavelength of the light in the visible spectrum). Light scattered from large particles follow the laws of reflection and refraction.

### 4.6.3 Application of Diffusers

If a laser is used as the illuminating light source for an imaging system, speckle noise from laser light can significantly deteriorate the image quality. A technique to overcome this issue is to use a rotating diffuser in the light path. This will cause the speckled pattern to move and the image to appear more uniform as opposed to a grainy image when no diffuser is used. When using a single rotating diffuser, the speckles will appear to be moving in one direction (this will be apparent when rotating the diffuser at slow speed). Adding a secondary stationary diffuser after the rotating diffuser will make the fringe appear to move in random directions. When the diffuser rotates at a speed such that the fringes move at rate much faster than the eye perception (or the detection rate of the imaging system if electronic imaging is used), then the image will appear more smooth than if no diffuser was used.

   Another application of a diffuser is to make discrete light sources appear more uniform. This can be achieved by placing a diffuser between the discrete light source and the observer. The pitch of the discrete arrays, the diffusion angle, and distance between the light source and the diffuser determine the uniformity of the illumination. An example is shown in Figure 4.9.

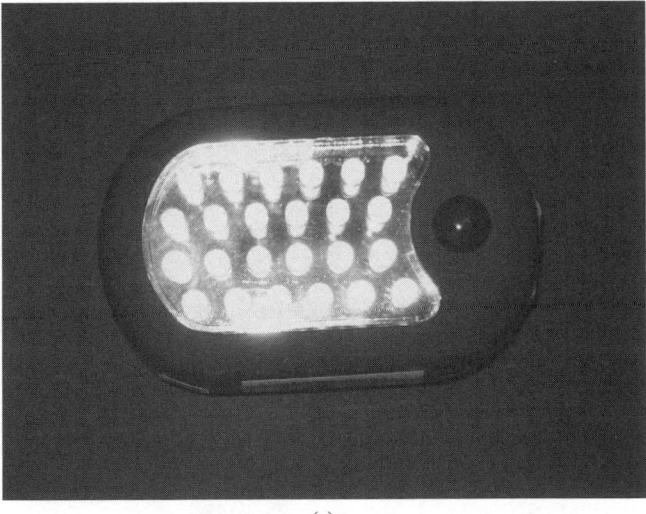

(a)

**FIGURE 4.9** Example of using a diffuser to generate uniform illumination from a discrete LED array. (a) LED array as light source (flash photograph). (b) LED array with no diffuser on. (c) Diffuser right on top of LED array. (d) Diffuser 3.5 cm from the surface of the LED array. (e) Diffuser 9.5 cm from the surface of the LED array. Diffuser used is a "wax paper." *(continued)*

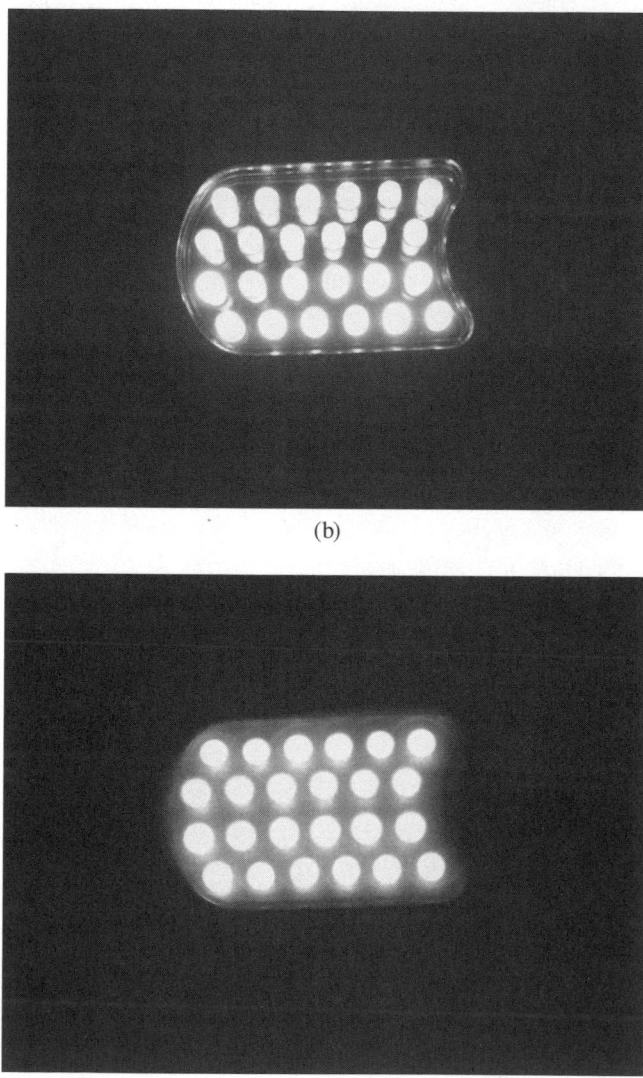

(b)

(c)

**FIGURE 4.9**  *(continued)* Example of using a diffuser to generate uniform illumination from a discrete LED array. (a) LED array as light source (flash photograph). (b) LED array with no diffuser on. (c) Diffuser right on top of LED array. (d) Diffuser 3.5 cm from the surface of the LED array. (e) Diffuser 9.5 cm from the surface of the LED array. Diffuser used is a "wax paper." *(continued)*

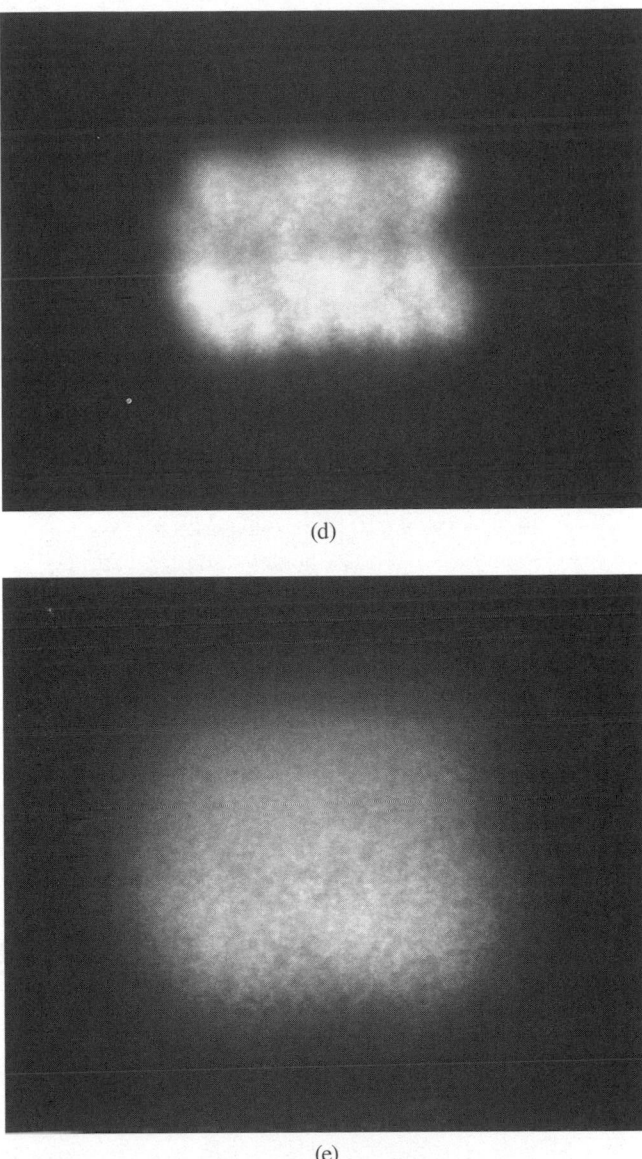

(d)

(e)

**FIGURE 4.9** *(continued)* Example of using a diffuser to generate uniform illumination from a discrete LED array. (a) LED array as light source (flash photograph). (b) LED array with no diffuser on. (c) Diffuser right on top of LED array. (d) Diffuser 3.5 cm from the surface of the LED array. (e) Diffuser 9.5 cm from the surface of the LED array. Diffuser used is a "wax paper."

## 4.7   SUGGESTED EXPERIMENTS

### EXPERIMENT 4.1: SINGLE AND DOUBLE SLIT
### (OR DOUBLE HOLE) DIFFRACTION

Purpose: To become familiar with diffraction concepts
Related simulations: Diffraction 1D
Materials needed:

- Laser pointer
- Aluminum foil or thick black paper
- Sharp to create the small holes or slits

#### EXPERIMENT

(Caution: Do not look directly into the beam, and do not look at the specular
reflection of the laser beam. Consult a laser safety manual [6,7] or laser safety
personnel before handling lasers).

1. Create a single and double holes or slits using the thin needle in the foil
   or the black paper. The hole(s) or slit width(s) should be smaller than the
   diameter of the laser spot, such that the laser spot will cover the entire slit
   or hole. If using a double slit or double hole, the laser spot should cover the
   both holes (or slits) (see illustration in Figure 4.10).
2. Illuminate the holes with a low-power laser, such as a laser pointer, as
   shown in Figure 4.11. View the diffracted beam using a screen (depending
   on the brightness of the spot, use either a white screen [e.g., white paper] or
   a dark screen [e.g., black paper]).
3. Observe the change in the diffraction pattern as you move the screen away
   from the slit.

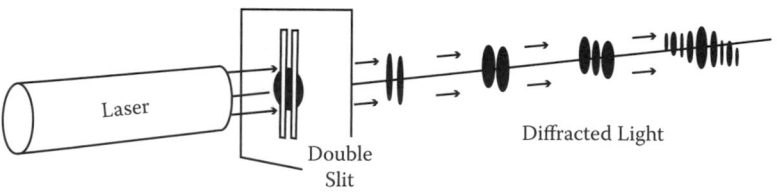

**FIGURE 4.10**   Double slit diffraction.

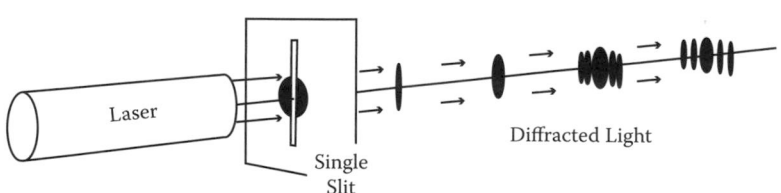

**FIGURE 4.11**   Single slit diffraction.

Some examples of sizes:
- Hole diameter: 0.45 mm; center to center separation: 0.75 mm
- Viewing distance: (a) 5 cm, (b) 35 cm, (c) 1.3 m
- Compare the observed diffraction pattern to Figure 4.12, Figure 4.13, and Figure 4.14 results for double slit diffraction examples.

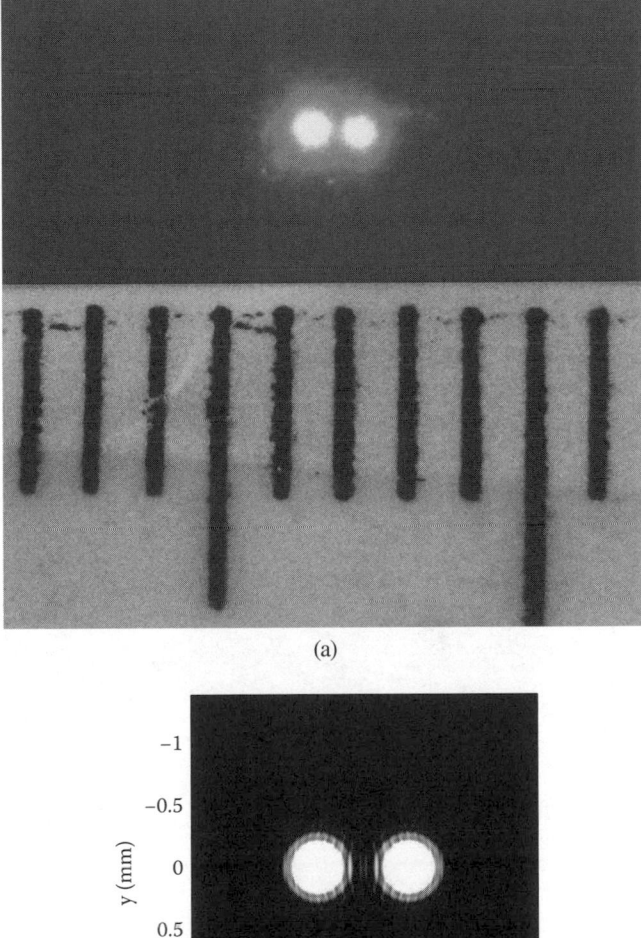

(a)

(b)

**FIGURE 4.12** Double pinhole diffraction. (a) Diffracted pattern photograph at distance 44 mm. Scale: 1 mm between each lines. (b) Simulated diffracted image and (c) intensity plot at distance 49 mm from the slit. Hole half width = 0.222 mm, separation (center to center) = 0.733 mm. Light source used: diode laser, 650 nm wavelength. *(continued)*

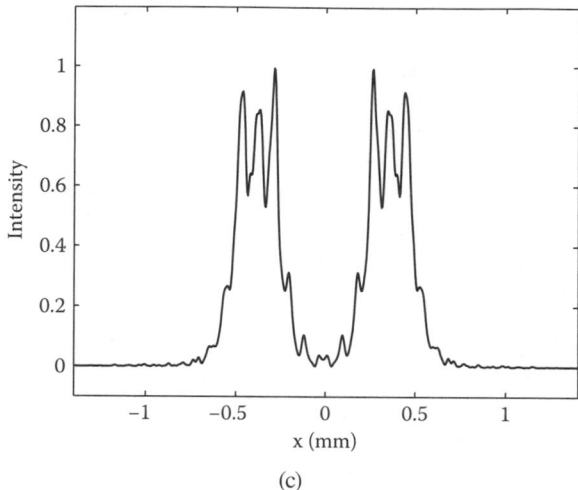

(c)

**FIGURE 4.12** *(continued)* Double pinhole diffraction. (a) Diffracted pattern photograph at distance 44 mm. Scale: 1 mm between each lines. (b) Simulated diffracted image and (c) intensity plot at distance 49 mm from the slit. Hole half width = 0.222 mm, separation (center to center) = 0.733 mm. Light source used: diode laser, 650 nm wavelength.

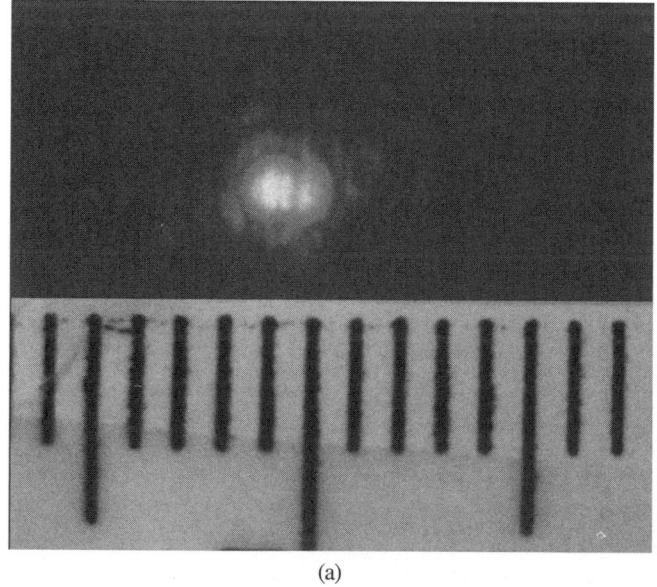

(a)

**FIGURE 4.13** Double pinhole diffraction. (a) Diffracted pattern photograph at distance 366 mm. Scale: 1 mm between each lines. (b) Simulated diffracted image and (c) intensity plot at distance 366 mm from the slit. Hole half width = 0.222 mm, separation (center to center) = 0.733 mm. Light source used: siode laser, 650 nm wavelength. *(continued)*

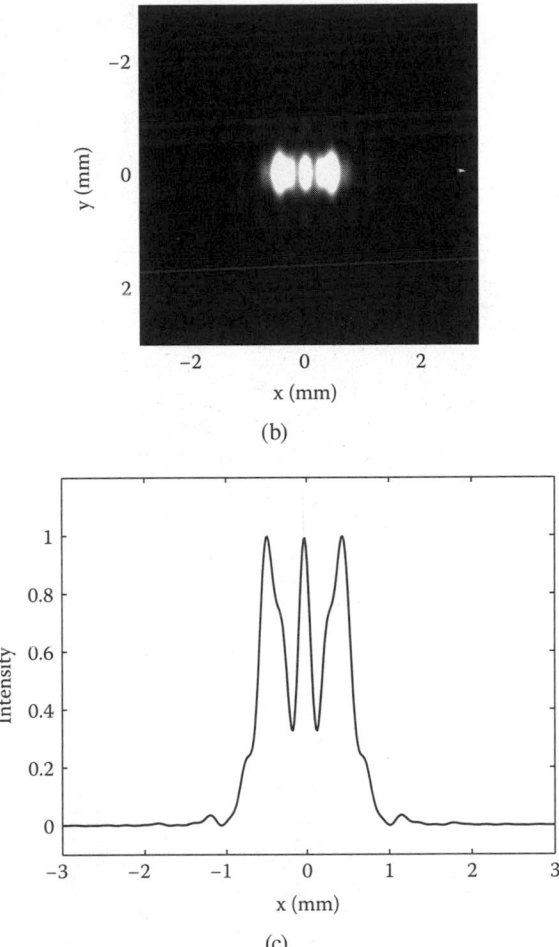

(b)

(c)

**FIGURE 4.13** *(continued)* Double pinhole diffraction. (a) Diffracted pattern photograph at distance 366 mm. Scale: 1 mm between each lines. (b) Simulated diffracted image and (c) intensity plot at distance 366 mm from the slit. Hole half width = 0.222 mm, separation (center to center) = 0.733 mm. Light source used: siode laser, 650 nm wavelength.

4. Optional. Compare the experimental findings to theoretical prediction using the single and double slit diffraction simulation. First estimate the size of the hole/slit and the separation. Enter these data as an input to simulation. Enter the wavelength (generally written on the laser [laser pointer]). If this information is not available, the wavelength of a red laser pointer would probably be between 633 nm and 670 nm. For example, entering 650 nm for a red laser would give close enough result.

5. Next enter the observation distance (see example). Compare the experimental result to the theoretical plots. If you are using a double holes aperture, then the diffraction pattern will be different in x-direction compared to y-direction. If the two holes are separated in the x-direction, then using a double slit simulation gives the intensity profile in the x-direction.

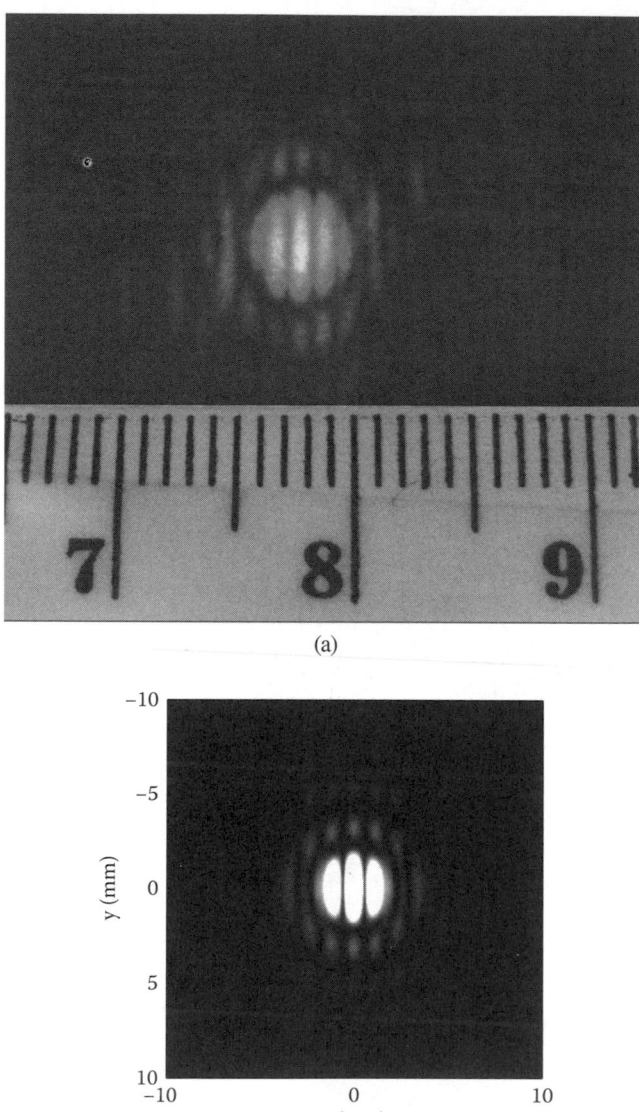

(a)

(b)

**FIGURE 4.14** Double pinhole diffraction. (a) Diffracted pattern photograph at distance 1295 mm. Scale: 1 mm between each lines. (b) Simulated diffracted image and (c) intensity plot at distance 1295 mm from the slit. Hole half width = 0.222 mm, separation (center to center) = 0.733 mm. Light source used: diode laser, 650 nm wavelength. Photograph and image in (a) and (b) are saturated to reveal details outside of the center spot. *(continued)*

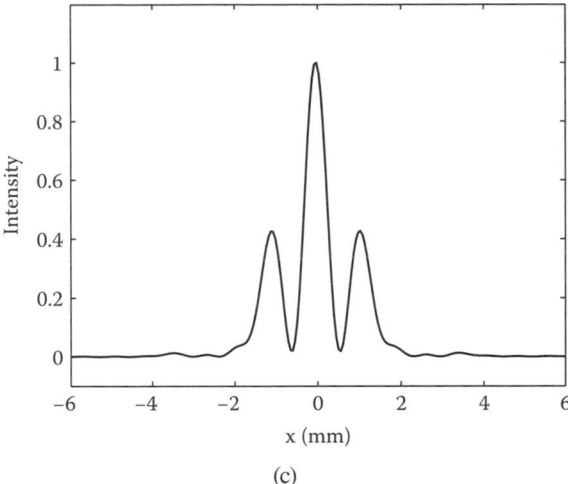

(c)

**FIGURE 4.14** *(continued)* Double pinhole diffraction. (a) Diffracted pattern photograph at distance 1295 mm. Scale: 1 mm between each lines. (b) Simulated diffracted image and (c) intensity plot at distance 1295 mm from the slit. Hole half width = 0.222 mm, separation (center to center) = 0.733 mm. Light source used: diode laser, 650 nm wavelength. Photograph and image in (a) and (b) are saturated to reveal details outside of the center spot.

Note the change in diffraction. The observed pattern is due to diffraction as light propagates in z direction. For example, for a double pinhole with diameter of 0.45 mm and center-to-center spacing of 0.75 mm, at z = 35 cm the observed pattern will be Fresnel diffraction. At a distance of 1.3 m, the observed pattern will be Fraunhofer diffraction, which resembles the Fourier transform of the slit.

### EXPERIMENT 4.2: INTERFERENCE FROM TWO SIDES OF GLASS

Purpose: To become familiar with interference concepts
Related simulations: N/A
Materials needed:

- Laser pointer
- Glass slide, such as a microscope slide (1 mm thick or thinner)
- Optional   Pointed heat source, such as a soldering iron. This will be used in the second part of the experiment to demonstrate fringe movement due to application of heat to one surface.

#### EXPERIMENT

(Caution: Do not look directly into the beam, and do not look at the specular reflection of the laser beam. Consult a laser safety manual [6,7] or laser safety personnel before handling lasers.)

1. Using the setup shown in Figure 4.15a, generate interference fringes from the front and back side of the glass interface. The purpose of the lens is to magnify the fringe image. An example of fringes generated with this setup is

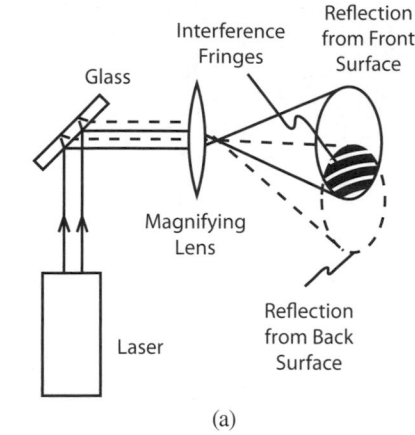

(a)

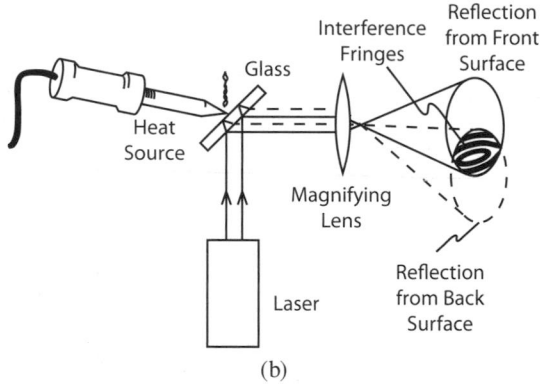

(b)

**FIGURE 4.15**   (a) Interference fringes from reflection of back and front side of a glass slide. (b) When heat is applied from a heat source (such as soldering iron), fringes move due to change due to localized change in the path length difference. This can be due to localized change in the shape of the glass, localized change in refractive index, or both.

shown in Figure 4.16a. The fringes are generated by interfering lift from the front and back surface of the glass.

2. As an optional experiment, use a heat source such as the tip of a soldering iron, and bring it close to the back surface of the glass. The effect of local heating of the glass will be observed from the fringe shape. For example, if the glass expands locally near the tip of the soldering iron, linear fringes will turn into circular fringes, as shown in Figure 4.15b. Examples of this phenomenon are shown in Figure 4.16, where the shape of the fringes shown in Figure 4.16a are altered when a soldering iron is brought close to the back surface of the glass, and as the soldering iron moved away from the surface of the glass, the fringe shape is changed as shown in Figure 4.16b,c,d.

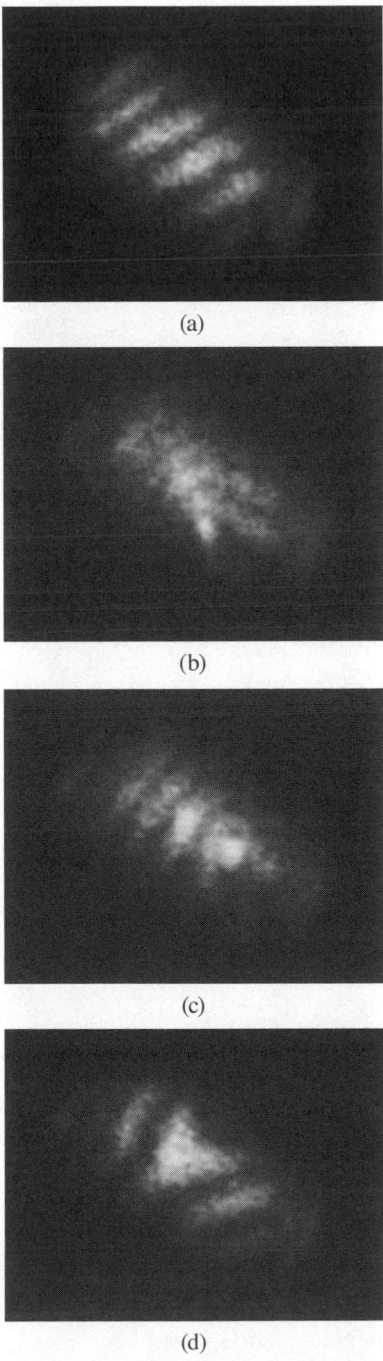

(a)

(b)

(c)

(d)

**FIGURE 4.16** Experimental observation of interference fringes described in Figure 4.16. (a) Interference fringes from reflection of back and front side of the glass. (b,c) Distorted fringes due to using a heat source, such as a soldering iron tip, near the back side of the glass. Light source: laser diode. Glass slab: 1 mm thick microscope slide.

It should be noted that lasers, and particularly laser pointers, vary greatly, and not all will produce desirable fringes. To achieve the best results, do the following.

a. Make sure there is good overlap of the spots reflected from the back and front sides of the glass.
b. If the interference fringes are not visible or are barely visible, then the laser pointer may have short coherence length (see discussion on coherence length in Chapter 2). In this case try another laser pointer. Alternatively, using a thin glass (e.g., 0.1 mm microscope slide instead of 1 mm slide) might solve this problem. The coherence length ($L_C$) of the laser should be larger than the optical path, namely

$$L_C > 2\ell \qquad\qquad (4.10)$$

or

$$L_C > 2\ell/\cos\theta_{in} \qquad\qquad (4.11)$$

where $\theta_{in}$ in the angle inside the glass.

## EXPERIMENT 4.3: RAYLEIGH SCATTERING WITH MILK AND WATER—COLOR CHANGE

Purpose: To familiarize with scattering
Related chapter/section: Manipulation of light/scattering
Related simulations: N/A
Materials needed:

- Glass container with flat sidewalls
- Halogen/incandescent (non-LED) flashlight (preferably dim)
- Milk and water
- Optional neutral density (gray) sunglasses for viewing.

### EXPERIMENT

(Caution: Do not stare directly into a bright flashlight. Use sunglasses if necessary. Do not use a very bright flashlight.)

When light encounters suspended particles that are small compared to the wavelength of the light, then scattering will be wavelength dependent. The scattered intensity is proportional to $1/\lambda^4$, where $\lambda$ is the wavelength of light. This is called Rayleigh scattering, and it is the reason why the sky is blue or why smoke appears bluish when illuminated with white light [3,4]. The aim of this experiment is to mimic small particle scattering phenomenon, which can be achieved by adding a small amount of milk.

The experimental setup is shown in Figure 4.17. Fill a glass container with water. Use an incandescent or halogen flashlight (not an LED flashlight) in the setup. Add a small amount of milk to the water. This will make the water murky. The amount of milk should be such that the filament of the flashlight should be barely visible when viewing light transmission through the water–milk mixture, as shown in Figure 4.17. Be sure not to use a very bright flashlight, and use dark

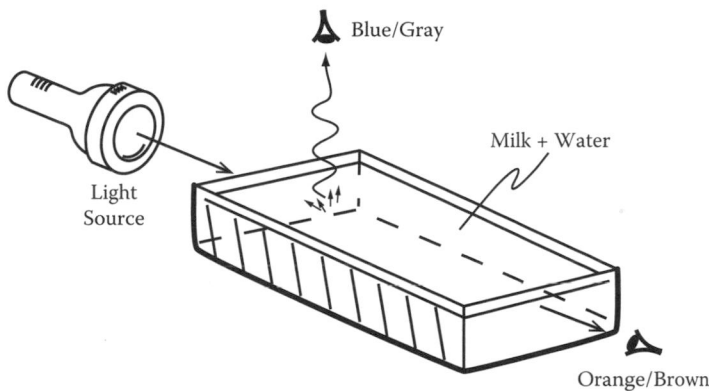

Blue/Gray

Milk + Water

Light
Source

Orange/Brown

**FIGURE 4.17** Experimental setup demonstrating wavelength dependence of scattering from small particles. A small amount of milk is added to the water in the glass container, just enough to make the water murky. Using an incandescent flashlight at the input will result blue-gray scattered light in the normal direction to the light propagation. Light through the mixture appear orange-brown, because scattered intensity is proportional to $1/\lambda^4$. Caution: Do not stare directly into bright light. Either use dark glasses or view at an angle.

neutral density (gray) sunglasses if necessary. Do not stare directly into the beam. Instead view at an angle.

Note the difference in color when viewing scattered light and transmitted light. The scattered light viewed from top (in the side of the container closer to the flashlight) looks blue-gray. In contrast, the transmitted light looks orange-brown. This is because scattered light intensity is proportional to $\lambda^{-4}$, and therefore shorter wavelengths (blue) scatter more than longer wavelengths (red-orange).

A continuation of this experiment is Experiment 11.5 in Chapter 11, where a similar setup is used for observing polarization effects of Rayleigh scattering from a milk–water mixture.

## REFERENCES

1. Goodman, J. W. 2004. *Introduction to Fourier Optics*. 3rd ed. Greenwood Village, CO: Roberts & Company.
2. Thelen, A. 1989. *Design of Optical Interference Coatings*. New York: McGraw-Hill.
3. Jenkins, F. A., and H. E. White. 1976. *Fundamentals of Optics*. 4th ed. New York: McGraw-Hill.
4. Born, M., and E. Wolf. 1999. *Principles of Optics: Electromagnetic Theory of Propagation, Interference and Diffraction of Light*. 7th ed. New York: Pergamon Press.
5. Kraft, P., C. Evangelista, M. Dacke, T. Labhart, and M. V. Srinivasan. 2011. Honeybee navigation: following routes using polarized-light cues. *Philosophical Transactions of the Royal Society B* 366:703–708.
6. Laser Institute of America. 2013. Laser safety information. www.lia.org/subscriptions/safet_bulletin/laser_safety_info.
7. Laser Institute of America. 2013. Laser pointer safety. www.lia.org/subscriptions/safet_bulletin/laser_pointer.

# 5 Polarization

Light consists of electric and magnetic fields, and the direction of the electric field is the direction of polarization. Light polarization can be linear, elliptical, or random. For a linear polarized light, the electric field oscillates in a single linear direction (e.g., up and down) as light propagates, as shown in Figure 5.1a. For a circular or elliptical polarized light, the electric field rotates as it propagates, as shown in Figure 5.1b. In most cases, light is randomly polarized, namely, the direction of polarization varies randomly in time as shown in Figure 5.1c.

## 5.1 POLARIZERS

Polarizers are optical components that only pass one polarization direction. Common uses of a polarizer are in photography and for polarized sunglasses. Their primary purpose is to minimize surface reflections. Polarizers are widely used in many sensing and communication applications.

When a polarizer is placed in the path of a randomly polarized light, then the output is linearly polarized. If a second polarizer is used, with polarization angle set at 90 degrees from the original of the first polarizer, then no light will pass through. If the first and the second polarizer are set to parallel, then light transmission is maximum. When a polarized light at angle $\theta = 0$ passes through a polarizer oriented at an arbitrary angle $\theta$, transmitted light intensity is given by

$$\frac{I(\theta)}{I(0)} = \cos^2 \theta \qquad (5.1)$$

where $I(0)$ is the intensity when the polarizer is set parallel to the incident polarization direction.

## 5.2 BIREFRINGENCE, RETARDATION, AND WAVE PLATES

Many materials exhibit anisotropic behavior, namely, the optical properties, such as the refractive index, is not equal in all directions. Therefore the material is said to exhibit birefringence. Examples are crystals, such as calcite and quartz. Noncrystalline materials include molded plastics or certain types of tape. Retarders, or wave plates, utilize birefringence properties of materials to alter the polarization state of an incident light. When light passes through a retarder, one polarization component is phase delayed relative to the other, and therefore the polarization state is altered.

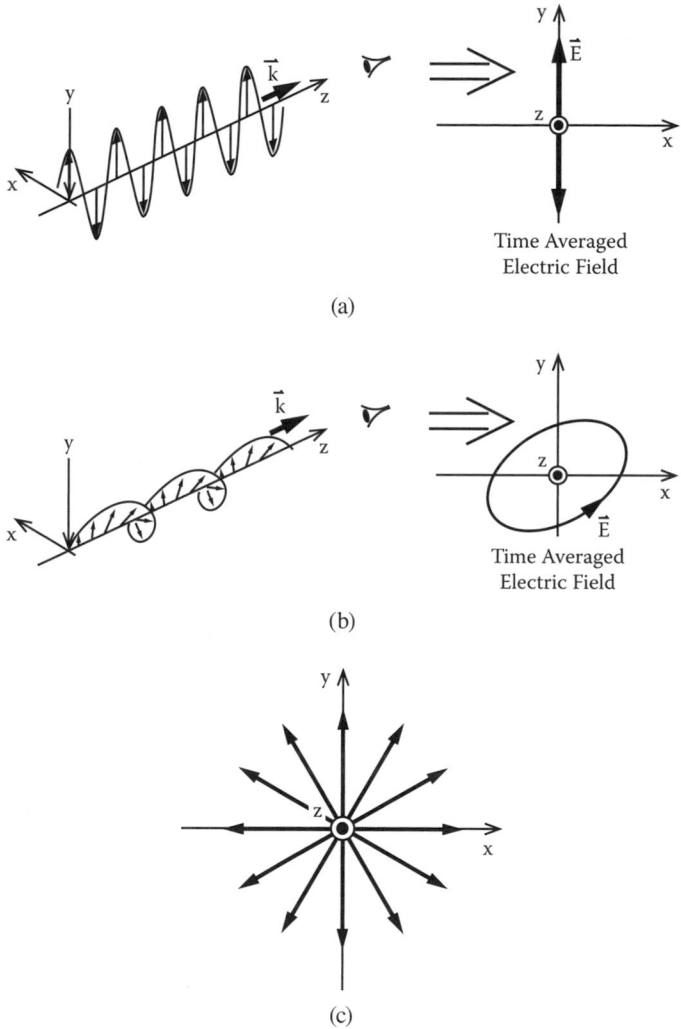

**FIGURE 5.1** (a) Linearly polarized light propagating vector k, and electric field E. (b) Elliptically polarized light. (c) Randomly polarized light where the electric field varies randomly in all directions.

Examples of some common retarders are quarter- and half-wave plates. A common use of a half-wave plate ($\lambda/2$ plate) is to rotate the polarization direction of a linear polarized light to another angle. For example, if the input polarization is horizontal, and if a half-wave plate is positioned at 45 degrees from horizontal, then light passing through the half-wave plate is polarized in the vertical direction.

A quarter-wave plate ($\lambda/4$ plate) is used to convert polarization from linear to elliptical, and vice versa. For example, if the input polarization is horizontal, and if

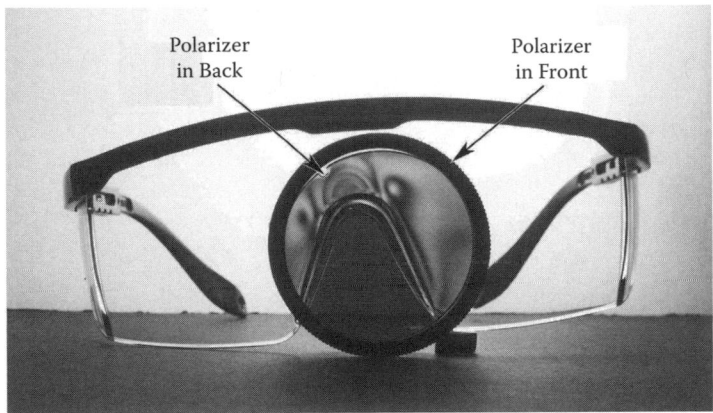

**FIGURE 5.2**   Example of birefringence in plastic goggles. The goggle is placed between two cross-polarizers. Fringes are due to varying birefringence, which is increasingly not uniform at the edges of the glass.

a quarter-wave plate is positioned at 45 degrees from horizontal, then light passing through the quarter-wave plate is circularly polarized.

When a wave plate is placed between two crossed polarizers, the transmitted light intensity depends on the orientation and retardation of the wave plate. Figure 5.2 shows such as an example, where a plastic goggle is placed between two cross-polarizers. Because of the varying birefringence in the plastic, the intensity is not uniform, and this is particularly apparent at the edges of the goggle. Such methods can be used to analyze birefringent materials and parts [1,2].

A relatively straightforward method of calculating polarization states is to use Jones vectors and matrices [1–3], where incident and transmitted polarization states are given by vectors, and the transformation matrix representing polarization component is given by a matrix. Appendix C describes Jones vectors and matrices.

Use of quarter- and half-wave plates, and the output polarizations can be visualized with the polarization MATLAB® simulation.

Suggested experiments included in this chapter demonstrate retardation concepts.

## 5.3   POLARIZED LIGHT REFLECTION

In Chapter 4, reflection coefficient of light at normal incidence was described. When light is incident on a surface at an angle, then reflection coefficients are polarization dependent and are different for two different polarization states. When light is parallel to the plane of incidence (the plane defined by the direction of propagation vector, $\bar{k}$), namely, it is parallel to the x-z plane shown in Figure 5.3a, then it is called p-polarized.

When the light is normal to the plane of incidence, namely, in y-direction as shown in Figure 5.3b, then it is called s-polarized.

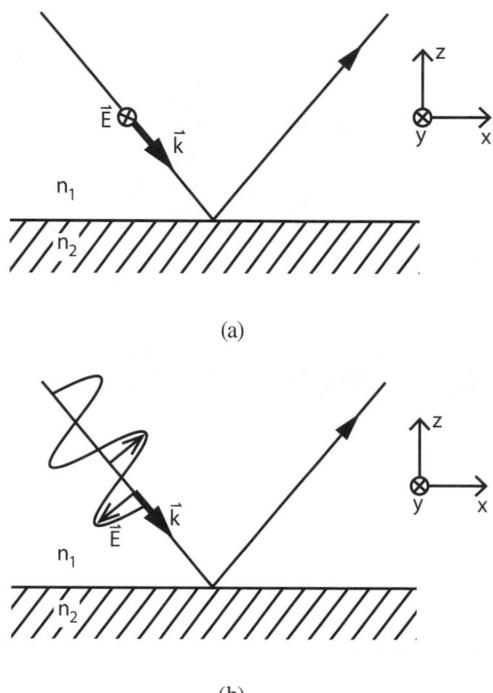

(a)

(b)

**FIGURE 5.3**   (a) s-polarized light. (b) p-polarized light. x-z is the plane of incidence.

The reflection coefficients and transmission of the s- and p-polarized lights for dielectric media are given by the Fresnel equations [1,4,5]:

$$r_s = \frac{n_1 \cos\theta_1 - n_2 \cos\theta_2}{n_1 \cos\theta_1 + n_2 \cos\theta_2} \tag{5.2}$$

$$t_s = \frac{2n_1 \cos\theta_1}{n_1 \cos\theta_1 + n_2 \cos\theta_2} \tag{5.3}$$

$$r_p = \frac{n_2 \cos\theta_1 - n_1 \cos\theta_2}{n_2 \cos\theta_1 + n_1 \cos\theta_2} \tag{5.4}$$

where $\theta_1$ is the incident angle in the first medium, and $\theta_2$ is the refracted angle in the second medium, calculated by Snell's law (described in Chapter 4). The reflected power ($R$) is obtained by

$$R = |r|^2 \tag{5.5}$$

At a particular angle of incidence only s-polarized light is reflected. This is called Brewster's angle ($\theta_B$) and is given by:

$$\theta_B = \tan^{-1}\left(\frac{n_2}{n_1}\right) \tag{5.6}$$

This phenomenon is used to minimize surface reflections, and it is the principle behind why polarized glasses reduce surface reflection. An example of using polarizers to minimize surface reflection is shown in Figure 5.4. In Figure 5.4a, the

(a)

(b)

**FIGURE 5.4** Minimizing surface reflection using polarizers. (a) Vase under a glass table viewed barely visible due to high reflections from the tabletop surface. (b) Vase under a table clearly visible when a polarizer is used in front of a camera set at vertical orientation. Vertical polarizer significantly reduces surface reflection.

surface reflection of a glass table is so bright that the vase under the table is not very visible. When a vertical polarizer is used in front of the camera, namely, blocking the s-polarization, then the surface reflection is minimized, and the vase under the table is clearly visible, as shown in Figure 5.4b.

There is also a difference between polarized light reflecting from a smooth and rough surface. For a smooth surface, such as glass, water, or tape, the reflected light polarization depends on the angle of incidence. This difference can be used to distinguish between rough and smooth surfaces. An example of this is shown in Figure 5.5. When tape is pasted on a rough wood surface, the tape is barely visible by an ordinary camera image. (Here the illumination and camera are at opposite from each other, both at approximately 50 degrees from the surface.) When a polarizer is used in front of the camera, then the tape becomes visible because of the difference in reflection amplitude between the tape and wood surfaces. If we take this one step further, and perform image difference calculation, namely, take the difference between the images without (Figure 5.5a) and with (Figure 5.5b) polarizer, then the resulting image (Figure 5.5c) clearly shows the shape and position of the area with the transparent tape. Another interesting observation in Figure 5.5b,c is the line at the lower left part of the image that becomes visible when a polarizer is used. This is a lightly scratched area, where locally, the wood grains have been smoothed and result in a difference in reflection amplitude.

(a)

**FIGURE 5.5** Demonstration of how reflection from two different types of surfaces have different types of reflection characteristics. A diffuse tape is on a wood board. (a) Tape not visible when viewed in a diffuse lighting. (b) A polarizer is used in front of the camera, and the tape becomes apparent. (c) Subtracting image (a) from (b), and after adjusting the contrast, the tape is revealed clearly. Light source and camera are apposite from each other, set at 50 degree from horizontal. *(continued)*

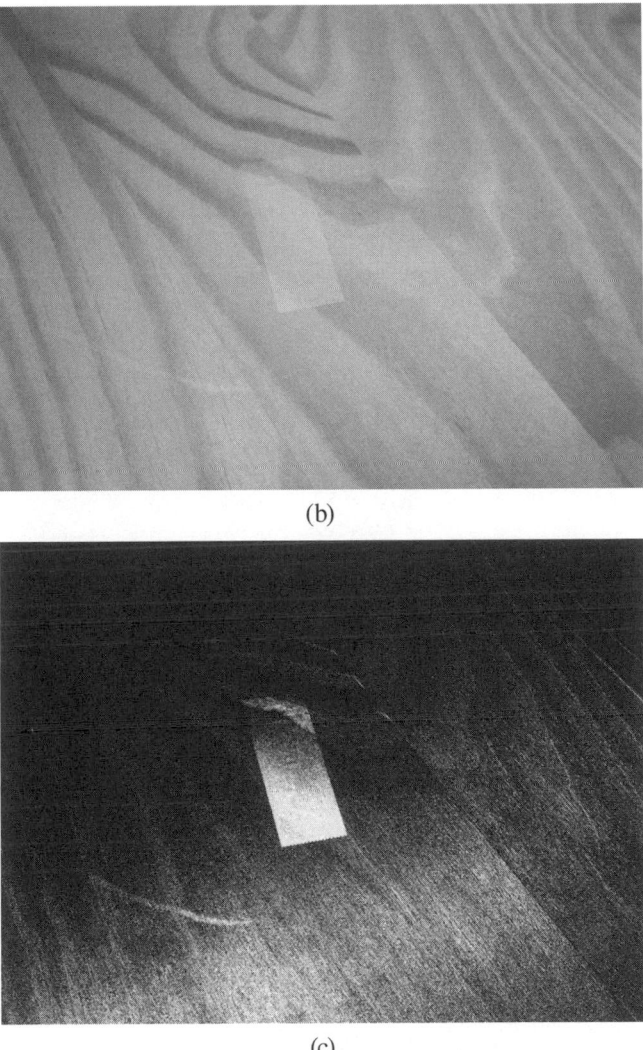

(b)

(c)

**FIGURE 5.5** *(continued)* Demonstration of how reflection from two different types of surfaces have different types of reflection characteristics. A diffuse tape is on a wood board. (a) Tape not visible when viewed in a diffuse lighting. (b) A polarizer is used in front of the camera, and the tape becomes apparent. (c) Subtracting image (a) from (b), and after adjusting the contrast, the tape is revealed clearly. Light source and camera are apposite from each other, set at 50 degree from horizontal.

## 5.4 POLARIZATION OF SMALL PARTICLE SCATTERING

To understand the scattering of light from small particles, the scattering molecule can be viewed as a dipole antenna [1]. The incident light makes the molecule oscillate in the direction of the incident light polarization. The molecule then oscillates in that direction and "radiates" as a dipole antenna at a specific direction.

As shown in Figure 5.6a, if the input light propagating in the z-direction is polarized vertically (in y-direction), then the molecule oscillates in the vertical (y) direction, and light is scattered in the x-direction, polarized vertically (y-polarized).
Namely, this is the direction and polarization of the dipole emission. However the
"dipole" does not radiate light that propagates in the y-direction.

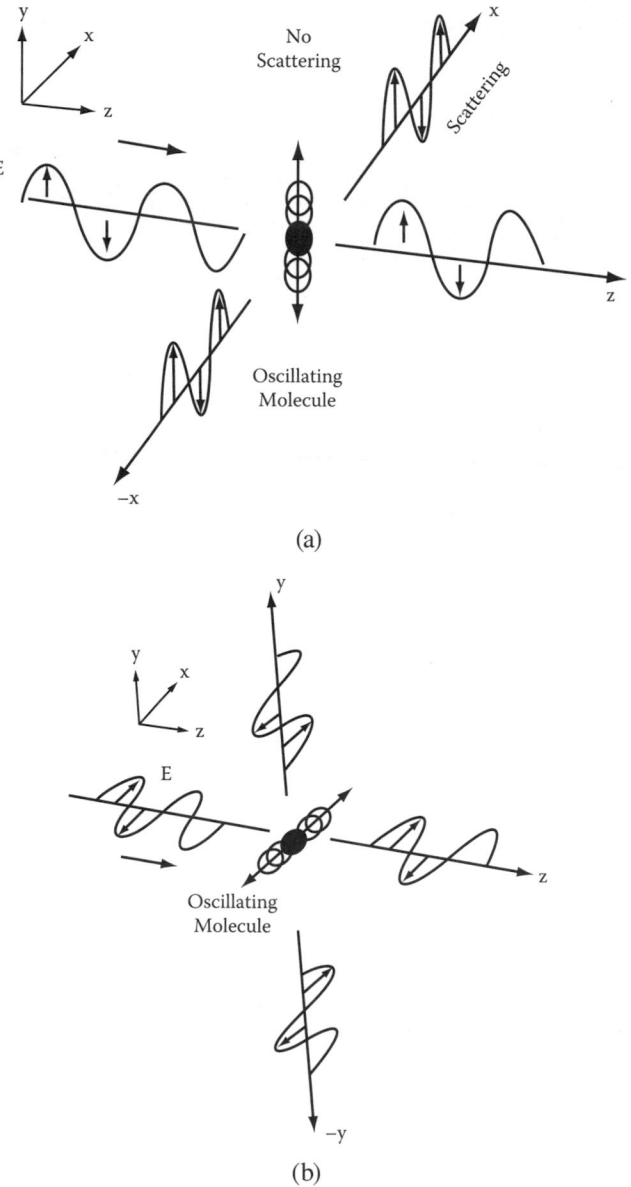

FIGURE 5.6  Polarization direction of scattering of light by a molecule (small particle compared to wavelength of light). (a) Incident light polarized in the x-direction. (b) Incident light
polarized in the y-direction. (c) Incident light is randomly polarized. *(continued)*

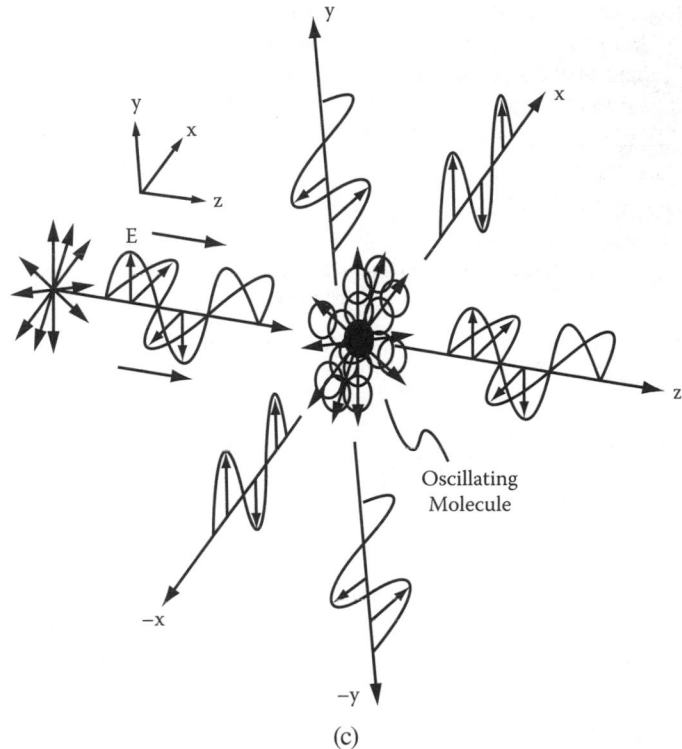

(c)

**FIGURE 5.6** *(continued)* Polarization direction of scattering of light by a molecule (small particle compared to wavelength of light). (a) Incident light polarized in the x-direction. (b) Incident light polarized in the y-direction. (c) Incident light is randomly polarized.

Similarly, when incident light is polarized in the x-direction (Figure 5.6b), then scattered light is propagating in the y-direction is also polarized in the x-direction.

When the incident light is randomly polarized, namely, it contains polarization components both in x- and y-directions (Figure 5.6c), then scattered light propagating in y-direction is polarized in x-direction, whereas light scattered in x-direction is polarized in y-direction, as shown in Figure 5.6c.

## 5.5 SUGGESTED EXPERIMENTS

### EXPERIMENT 5.1: FINDING POLARIZER ANGLE

Purpose: A quick method of finding polarizer axis
Related simulations: N/A
Materials needed:

- One polarizer or one set of polarized sunglasses
- Flat black plastic or metal surface (e.g., car dashboard, any black asphalt road surface)

### EXPERIMENT

1. View the black-colored surface such that the light source is (at a shallow angle such as 40 degrees or less). Without using a polarizer, you will see some reflection at the surface.
2. Place the polarizer in front of your eye, and rotate it ±90 degrees. At a particular angle, the surface will look matte; namely the surface reflection will be minimized. This angle is when the polarizer angle is normal to the surface. If the surface is horizontal, the polarizer angle is vertical.

### EXPERIMENT 5.2: POLARIZATION AND BIREFRINGENCE EXPERIMENTS, PART 1

Purpose: To understand polarization and birefringence
Related simulations: Polarization simulations
Materials needed:

- Light source: Flashlight or simply room light scattered from a white paper
- Two polarizers (can be purchased from camera store or optical supplies store); alternative, use two polarized sunglasses
- Piece of plastic, such as clear part of the compact disk (CD) case or clear packaging plastic

### EXPERIMENT

Orient the polarizers at 90 degrees from each other. When the polarizers (sometimes referred to as polarizer analyzer set) are at 90 degrees from each other, light transmission will be at minimum. Insert the plastic piece between the two crossed-polarizers and note that light will pass again. Note that different parts of the film will have different transmission. In a plastic piece, color fringes will be observed, indicating varying birefringence across the plastic part. An example of this is shown in Figure 5.2, where a birefringence in the plastic goggle is apparent when placed between two crossed-polarizers.

### EXPERIMENT 5.3: POLARIZATION AND BIREFRINGENCE EXPERIMENTS, PART 2

Purpose: To understand polarization and birefringence
Related simulations: Polarization simulations
Materials needed:

- Light source: Flashlight or simply room light scattered from a white paper
- Two polarizers (can be purchased from camera store or optical supplies store); alternative, use two polarized sunglasses
- Half-wave plate (from optics supply stores, or see Experiment 5.4 on how to make a wave plate)
- Quarter-wave plate (from optics supply stores, or see Experiment 5.4 on how to make a wave plate)

### EXPERIMENT

This is a continuation of the previous experiment (Experiment 5.2) but uses more controlled components, such as quarter- and half-wave plates.

1. Orient the polarizers at 90 degrees from each other, then insert a half-wave plate ($\lambda/2$ plate) in between. Rotate the wave plate. When the wave plate

axis is either parallel to the first or the second polarizer, then light transmission will be nearly zero. When the wave plate is oriented at 45 degrees with respect to one of the polarizers, then transmission is at maximum. These findings can be verified by the MATLAB® *Polarization Simulation*, by setting P1 angle to 0, P2 angle to 90°, and wave plate angle to 45° or 135°.

2. Orient the polarizers parallel to each other, then insert a half-wave plate ($\lambda/2$ plate) in between. Rotate the wave plate. When the wave plate axis is parallel to the polarizers, then light transmission will be maximum. When the wave plate is oriented at 45 degrees with respect to the polarizers, then light transmission is zero (or minimum), because light polarization was rotated 90 degrees by the half-wave plate and does not transmit through the second polarizer. These findings can be verified by the MATLAB® *Polarization Simulation*, by setting P1 and P2 angles to 0, and set the wave plate angle to 0° and 45° and noting the change in $I_{Rel}$.

3. Orient the polarizers at 90 degrees from each other, then insert a quarter-wave plate ($\lambda/2$ plate) in between. Rotate the wave plate. When the wave plate axis is parallel to the first or the second polarizer, then light transmission will be nearly zero. When the wave plate is oriented at 45 degrees with respect to one of the polarizers, then transmission is partial. In fact transmission will be 50% of what it would be if the two polarizers and the quarter-wave plate axes were parallel. These findings can be verified by the MATLAB® *Polarization Simulation*, by setting the P1 angle to 0, and the P2 angle to 90°. When the wave plate angle is set to 45° or 135°, $I_{Rel} = 0.5$.

### EXPERIMENT 5.4: HOW TO MAKE A WAVE PLATE

Purpose: To understand polarization and birefringence
Related simulations: Polarization simulation
Materials needed:

- Transparent (clear) tape
- Piece of glass
- Two polarizers

#### EXPERIMENT

It should be noted that this experiment works only if the tape is birefringent. If one type of tape does not work, experiment with another brand of tape.

1. Create a wave plate by placing a layer of tape on the glass.
2. On the same orientation (e.g., in x-direction, see Figure 5.7) and on the top of the tape, add another layer of tape but intentionally offset it such that when viewed between two polarizers, one can see the effect of retardation. Use up to five or six layers of tape.
3. Find the number of layers that correspond to a half-wave ($\lambda/2$) and quarter-wave ($\lambda/4$) plates. Hold the tape–glass piece (namely the wave plate) between two cross-polarizers and rotate until you find the brightest area (see Figure 5.8). The number of layers that give brightest transmission is closest to half-wave plate. The layer that is approximately half the maximum of the layer with the maximum transmission is closest to quarter-wave ($\lambda/4$) plate. Note the angle of orientation of the wave plate with respect to the first polarizer. The area that corresponds to half-wave plate is at 45 degrees from the polarizers. Note which area corresponds to half-wave and quarter-wave plates.

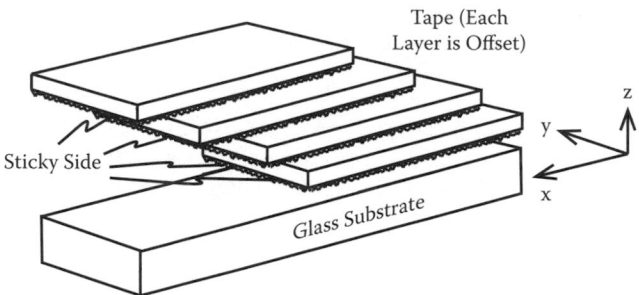

**FIGURE 5.7**   Fabricating a wave plate using transparent (clear) tape. Layers are offset to clearly see the additive effect of each layer.

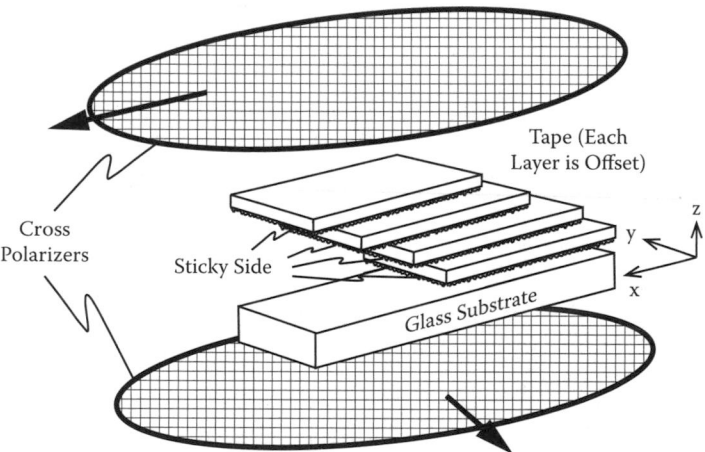

**FIGURE 5.8**   Testing with two cross-polarizers to find the number of layers that correspond to quarter-wave plate.

4. To verify, rotate the second polarizer so that it is parallel to the first polarizer. The transmission without the wave plate is at maximum. Now insert the tape/glass piece in between the two parallel polarizes, and rotate, as shown in Figure 5.9. When the area that corresponds to quarter wave plate is at 45 degrees from either one of the polarizers, then the transmission will be maximum.

## NOTES

- Be aware that you should keep the wave plate normal to direction of light propagation. Tilting it will change the wave plate characteristic because it increases propagation distance through the tape by the cosine of the angle, and therefore increase the retardation.

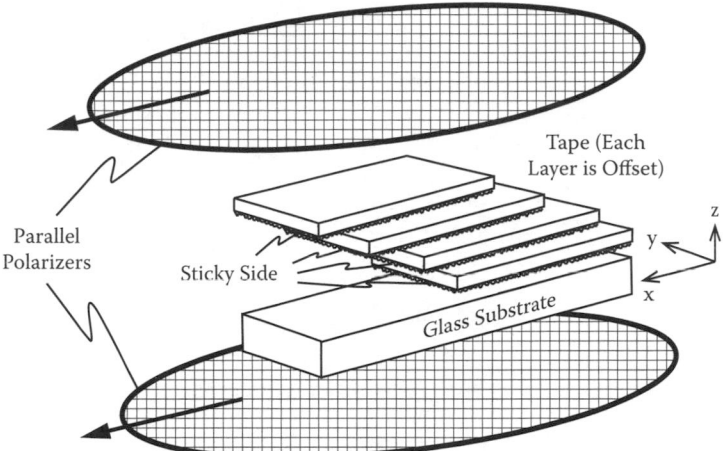

**FIGURE 5.9**  Testing with two parallel polarizers to find the number of layers that correspond to half-wave plate.

- When adding the tape layers, always add the tape in the same orientation, such as in x-direction (see Figure 5.10). If the direction is reversed, such as if one layer is in the x-direction and the other layer is in the y-direction, or if tape is applied on the opposite side of the glass (as shown in Figure 5.11), then the birefringence effect will cancel or minimized.
- An alternative method of making a wave plate is to use mica sheet (available from some hardware stores), and peel it layer by layer using an adhesive tape, while measuring the birefringence as described earlier (see Figure 5.8 and Figure 5.9).
- An example of five layers of tape placed between two cross-polarizers is shown Figure 5.12.

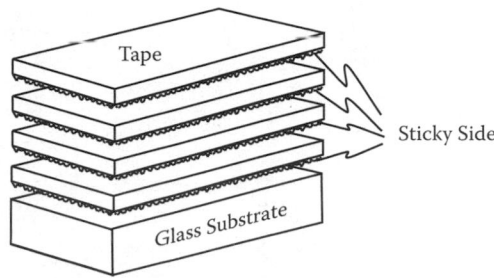

**FIGURE 5.10**  Correct way to use multiple layers of transparent tape to form a wave plate.

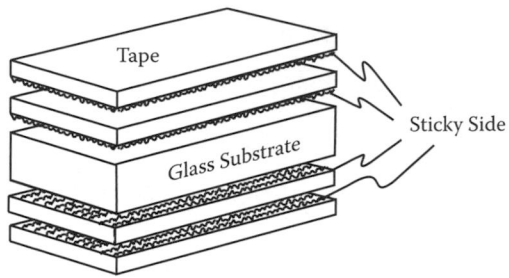

**FIGURE 5.11**    Incorrect way to use multiple layers of transparent tape to form a wave plate.

### EXPERIMENT 5.5: RAYLEIGH SCATTERING FROM THE SKY

Purpose: To familiarize with polarized scattering
Related chapter/section: Polarization
Related simulations: N/A
Materials needed: Polarized sunglasses or polarizer

#### Experiment

When light encounters suspended particles that are small compared to the wavelength of the light, then scattering is wavelength dependent. The scattered intensity is proportional to $1/\lambda^4$, where $\lambda$ is the wavelength of light. This is called Rayleigh scattering, and it is the reason why the sky is blue [3],[4].

     The best time to perform this experiment is when the sun is low in the horizon. Using polarized sunglasses, view the sky while the sun is toward the viewer's shoulder,

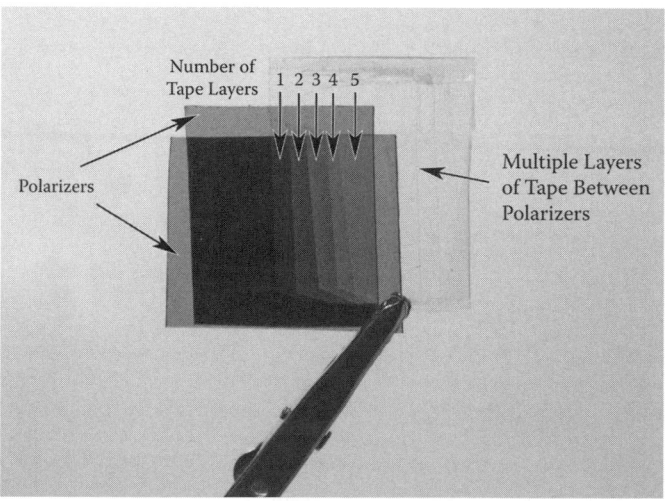

**FIGURE 5.12**    Demonstration of using multiple layers of transparent tape as a means of wave plate. Five layers of tape are sandwiched between two cross-polarizers. Each layer is offset, as shown in Figure 5.6, to demonstrate the additive effect of birefringence.

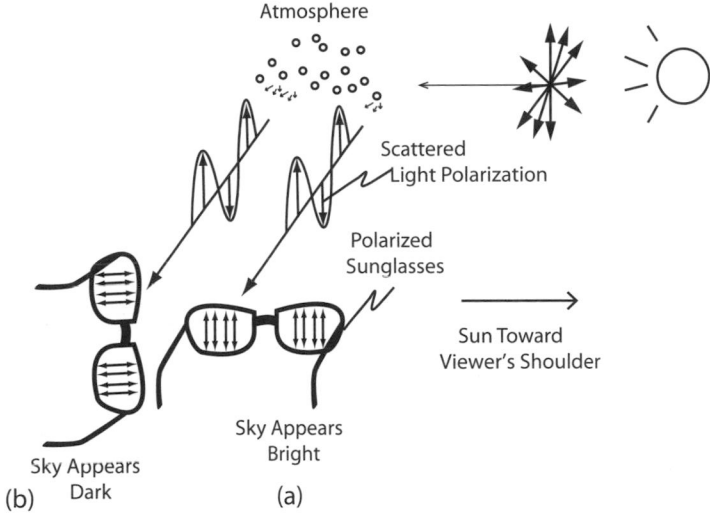

Atmosphere

Scattered
Light Polarization

Polarized
Sunglasses

Sun Toward
Viewer's Shoulder

Sky Appears
Bright

Sky Appears
(b) Dark (a)

**FIGURE 5.13** Observation of polarization direction of light scattered from the sky.

as shown in Figure 5.13a. When viewing upright, namely, when sunglasses are parallel to the horizon (polarization axis oriented vertical), the sky appears bright. If the viewer's head is tilted such that the sunglasses are perpendicular to the horizon (polarization axis oriented horizontal), then the sky appears dark, as shown in Figure 5.13b.

The same experiment can be repeated using a polarizer, such as a polarizer filter used for SLA cameras. When the polarizer is oriented such that polarization axis is vertical, the sky looks bright. When the polarizer is rotated such that the polarization axis is parallel to the horizon, then the sky looks dark.

The reason for the polarization direction is explained earlier in the chapter in the section titled "Polarization of Small Particle Scattering."

Compare the findings of this experiment to Experiment 11.5 in Chapter 11.

## REFERENCES

1. Hecht, E. 2001. *Optics.* 4th ed. Reading, MA: Addison-Wesley.
2. Kliger, D. S., J. W. Lewis, and C. E. Randall. 1990. *Polarized Light in Optics and Spectroscopy.* Boston: Harcourt Brace Jovanovich.
3. Yariv, A., and P. Yeh. 2002. *Optical Waves in Crystals: Propagation and Control of Laser Radiation.* New York: Wiley Interscience.
4. Born, M., and E. Wolf. 1999. *Principles of Optics: Electromagnetic Theory of Propagation, Interference and Diffraction of Light.* 7th ed. New York: Pergamon Press.
5. Jackson, J. D. 1998. *Classical Electrodynamics.* 3rd ed. New York: John Wiley & Sons.

# 6 Geometrical Optics

Geometrical optics treats all light as rays, to estimate light path, imaging, and optical properties such as aberration and light distribution. Since it deals with light rays, it is most suitable for optical systems where light can be represented as rays. Geometrical optics is used when typical dimensions are much larger than the wavelength light. In applications such as waveguides that are on the order of wavelength, treatment of light as waves, rather than rays, is more appropriate, and requires calculations that involve wave equations.

Even though many diffractive elements (such as gratings) deal with features that are on the order of light wavelength, it is still possible to combine refractive elements in geometrical optics calculations. As long as the overall dimension of diffractive elements is large enough, such as a diffractive grating with a diameter much larger than the wavelength of the light, then it can be treated as a bulk optical element and be combined in the optical system with the rest of the elements. In this case, only the macroeffects of the diffractive element are utilized.

## 6.1   RAY TRACING METHODS

Ray tracing is simply performed by following the refraction and reflection principles described in Chapter 4. For example, for a refractive elements such as a thick lens shown in Figure 4.7 (Chapter 4), at every point on the glass–air interface, light follows the same refraction rules as indicated in inset of Figure 4.7. Basically the laws of refraction governed by Snell's law are simply applied at every interface. All one has to do is find the surface normal. The rest of ray tracing puts all the rays into a geometrical context.

The same applies to reflective interfaces, where laws of reflection apply, as shown in Figure 4.4 (Chapter 4). As shown in the inset, incident angle is equal to reflection angle, all measure with respect to surface normal.

After establishing ray tracing, the next step is to measure other characteristics, such as aberrations, which is a measure of how well an optical system has the ability to image an object.

To perform a rough ray tracing design of an optical system, a simple ray tracing is a good starting point. Several options exist to perform this task, such as:

- Graphical method
- Thin lens formulation
- Ray tracing with refraction equations utilizing spreadsheet programs
- Principal plane method
- Matrix method

After performing an approximate design of the optical system, more detailed analysis (if needed) and optimization can be performed with commercial optical design software.

## 6.2 LENS MAKER'S FORMULA

The effective focal length of a thick lens surrounded by air is [1]

$$\frac{1}{f} = (n-1)\left[\frac{1}{R_1} - \frac{1}{R_2} + \frac{(n-1)d}{nR_1R_2}\right] \tag{6.1}$$

where $R_1$ and $R_2$ are the radii of curvature of the first and second surface, $d$ is the lens thickness, and $n$ is the refractive index of the lens, as shown in Figure 6.2a.

## 6.3 THIN LENS FORMULATION

For a thin lens ($d \rightarrow 0$) Equation (6.1) becomes

$$\frac{1}{f} = (n-1)\left[\frac{1}{R_1} - \frac{1}{R_2}\right] \tag{6.2}$$

This is a very useful technique for calculating image positions. Here, all the lenses are treated as if they are infinitely thin. The relation between object and image distance with respect to the lens focal length is

$$\frac{1}{f} = \frac{1}{s_1} + \frac{1}{s_2} \tag{6.3}$$

where $f$ is the focal length of the lens, and $s_1$ and $s_2$ refer to object and image, respectively. Similarly, the transverse magnification is

$$M_T = -\frac{s_2}{s_1} \tag{6.4}$$

When $M$ is negative, it means the image is flipped. When $M$ is more than 1, then there the image is larger than the object (magnified). When $M$ is less than 1, the image is smaller than the object (demagnified).

This system follows the optical sign convention described in Table 6.1.

For a multiple lens system, Equation (6.3) can be repeated. For example, when two cascaded lenses are used, after calculating $s_2$ for the first lens, and knowing the distance between the first and the second lens, then a new $s_1$ is calculated. Equation (6.3) is then used to calculate $s_2$ for the second lens. This process is repeated for all subsequent optical components.

## 6.4  RAY TRACING FORMULATION

Ray tracing involves utilizing reflective (Equation 4.1) and refractive (Equation 4.2) ray equations (see in Chapter 4). The calculation can be setup in a spreadsheet program. Several such methods are detailed by Kingslake [2], such as the *y-nu* method of ray tracing, which can be achieved using a spreadsheet program. There are a number of methods that simplify the ray tracing calculations, so that a rough estimate of image location can be obtained or a preliminary optical design can be achieved.

## 6.5  GRAPHICAL METHOD OF RAY TRACING

The graphical method of ray tracing involves following the light path graphically, using laws such as reflection (Equation 4.1 in Chapter 4), refraction (Equation 4.2 in Chapter 4), or thin lens formulation (Equation 6.3). In many instances this method can be a rapid technique to obtain imaging parameters. An image formed by a positive thin lens is shown in Figure 6.1. In Figure 6.1a, the position and height of the

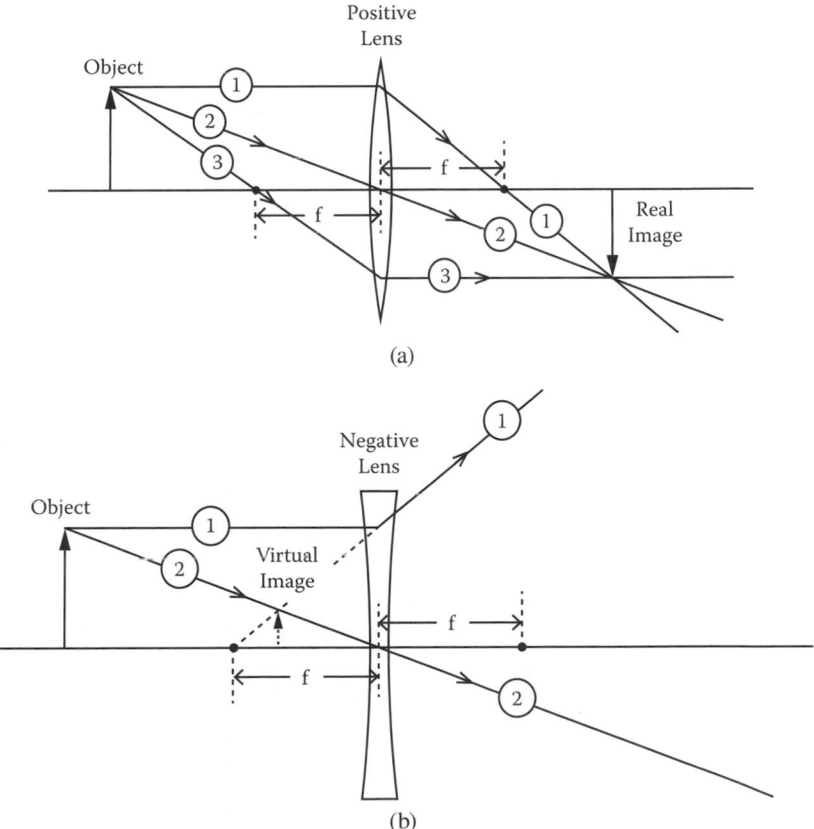

**FIGURE 6.1**  Graphical method of finding a location of image for a (a) positive lens and (b) a negative lens.

image of a positive lens can be found using the intersection of two rays (rays 1 and 2, 2 and 3, or 1 and 3). Similarly, in Figure 6.1b, the position and height of the image of the virtual image from a negative lens can be found using the intersection of two rays, 1 and 2.

In some instances, for example, when multiple optical components are involved, the graphical method may prove too cumbersome. However it can be combined with simple thin lens formula or calculation methods to obtain imaging parameters.

## 6.6   PRINCIPAL PLANE METHOD

When optical systems consist of a number of thick lens components, thin lens formulation no longer applies. However, using *principal planes*, it is possible to use thin lens formulation. The procedure is as follows.

1. Find the principal planes of the thick lens or optical component using the following formulation [1]:

$$V_1 H_1 = \frac{-f(n-1)d}{nR_2} \tag{6.5}$$

$$V_2 H_2 = \frac{-f(n-1)d}{nR_1} \tag{6.6}$$

where $f$ is the effective focal length, calculated from Equation (6.1); and $R_1$ and $R_2$ are the lens radii, as shown in Figure 6.2a. $V_1 H_1$ and $V_2 H_2$ are the distance between the lens vertices ($V_{1,2}$) and principal points ($H_{1,2}$), the point where principal planes cross the optical axis. The focal point of the lens is shown in Figure 6.2b.

2. Use thin lens formulation, find object and image distances, using Equation (6.3), as shown in Figure 6.2c.

Repeat the same for the subsequent components and distances. For example, if another lens is added after the first lens, do the following: (1) Calculate the effective focal length of the second lens ($f_{Lens2}$). (2) Calculate the principal planes for the second lens. (3) Find $s_{1,Lenw2}$ for the second lens, which is the distance from the *image position of the first lens* to the *first principal plane of the second lens*. (4) Calculate $s_{2,Lens2}$ from Equation (6.3) using $s_{1,Lenw2}$ and with effective focal length of the second lens $f_{Lens2}$. (5) Calculate the new image position, which is at a distance $s_{2,Lens2}$ measured from the second principal plane of the second lens. While doing these calculations, keep track of the sign convention for distances and radii.

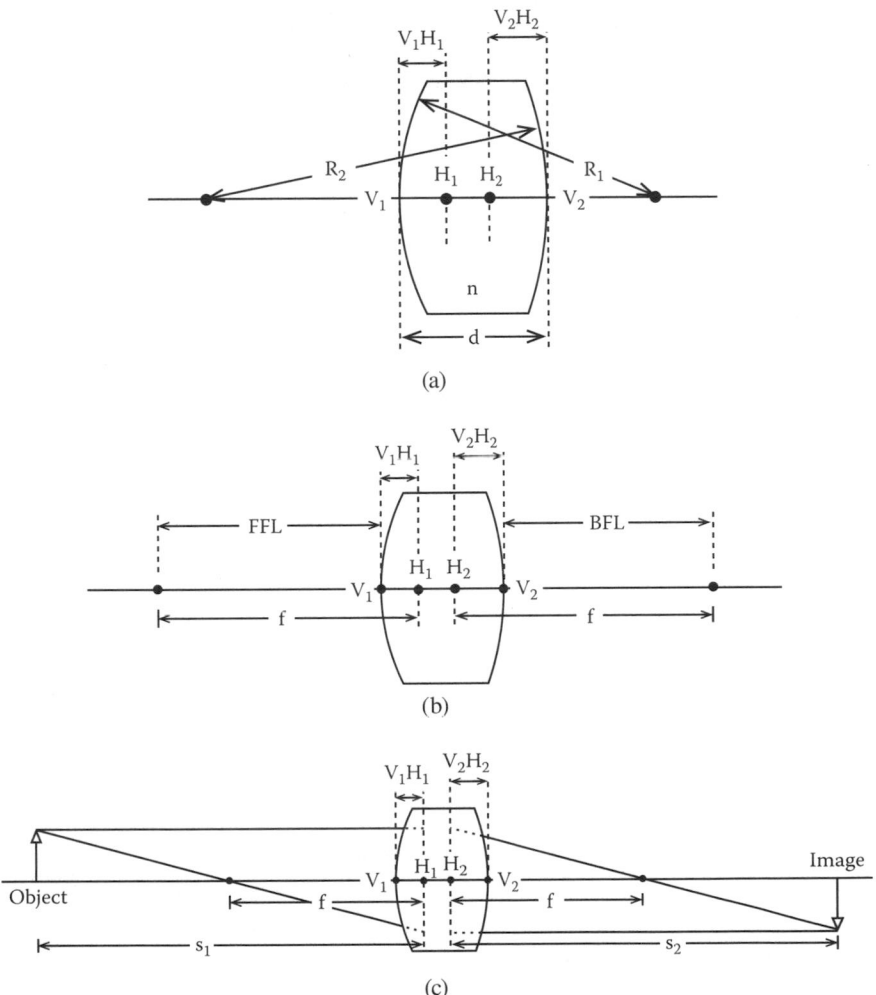

**FIGURE 6.2** (a) Thick lens and principal planes. (b) Principal planes and lens radii. (c) Effective focal length distances, and position of front focal length (FFL), and back focal length (BFL). (d) Ray tracing with principal planes and finding image position using principal planes.

## 6.7 MATRIX METHOD OF RAY TRACING

Matrix method is another method of ray tracing, which is particularly useful for using with software capable of matrix calculations (such as MATLAB®). The optical system is represented by a matrix, and the input ray and the output ray is represented by a vector, namely,

$$
\begin{bmatrix} n'\alpha' \\ y' \end{bmatrix} = \begin{bmatrix} M_{11} & M_{12} \\ M_{21} & M_{22} \end{bmatrix} \begin{bmatrix} n\alpha \\ y \end{bmatrix} \tag{6.7}
$$

where $n$, $\alpha$, and $y$ refer to the input refractive index, ray angle, and ray height at a given point; and $n'$, $\alpha'$, $y'$, refer to the output refractive index, ray angle, and ray height, respectively, after passing through the optical system represented by the matrix $\bar{M}$ with elements $M_{ij}$.

Each optical component, surface, or propagation can be represented in a matrix, and multiple elements can be combined into one matrix by multiplying each matrix that is representative of the other. When multiplying transfer matrices 1 to $N$, the system matrix is given by

$$\ddot{M}_{Tot} = \ddot{M}_N \ddot{M}_{N-1} \cdots \ddot{M}_3 \ddot{M}_2 \ddot{M}_1 \tag{6.8}$$

The system matrix can be used to calculate output parameters $n'$, $\alpha'$, $y'$ by multiplying the input vector by the system matrix

$$\begin{bmatrix} n'\alpha' \\ y' \end{bmatrix} = \ddot{M}_{Tot} \begin{bmatrix} n\alpha \\ y \end{bmatrix} \tag{6.9}$$

Some of the matrix representations are given in Table 6.2, with a pictorial representation and sign convention. The formulation is based on small angle approximation. Namely, the Snell's law, $n_1 \sin \theta_1 = n_2 \sin \theta_2$ (described in Chapter 4) is replaced by $n_1 \cdot \theta_1 = n_2 \cdot \theta_2$.

## TABLE 6.1
## Sign Convention

| Parameter | Sign | Condition or Refer To |
|---|---|---|
| Radius ($R$) | + | C is to the right of vertex ($V$) |
| Radius ($R$) | − | C is to the left of vertex ($V$) |
| Object distance ($s_1$) | + | Object to the left of lens |
| Object distance ($s_1$) | − | Object to the right of lens |
| Image distance ($s_2$) | + | Image to the right of lens (real image for a single lens) |
| Image distance ($s_2$) | − | Image to the left of lens (virtual image for a single lens) |
| Object and image height ($y$) | + | Object or image above optical axis |
| Focal length ($f$) | + | Converging lens, also referred to as positive lens |
| Focal length ($f$) | − | Diverging lens, also referred to as negative lens |
| Transverse magnification ($M_T$) | + | Erect image |
| Transverse magnification ($M_T$) | − | Inverted image |
| Principal plane distances from vertex ($V_1H_1$), ($V_2H_2$) | + | $H$ to the right of vertex ($V$) |
| Principal plane distances from vertex ($V_1H_1$), ($V_2H_2$) | − | $H$ to the left of vertex ($V$) |

The matrix method and matrix describing optical elements and systems are discussed in detail by Hecht [1], Klein and Furtak [3], and Mahajan [4].

## 6.8  SIGN CONVENTIONS

To perform calculations for thin lenses, thick lenses, and principal plane formulae, a sign convention that is based on the assumption that light traveling from left to right is given in Table 6.1. Sign convention for radii of refractive lenses is shown in Figure 6.3.

An example for calculating the effective focal length of a double convex lens (similar to that illustrated in Figure 6.2a) using the sign convention of Table 6.1 is shown in Table 6.3.

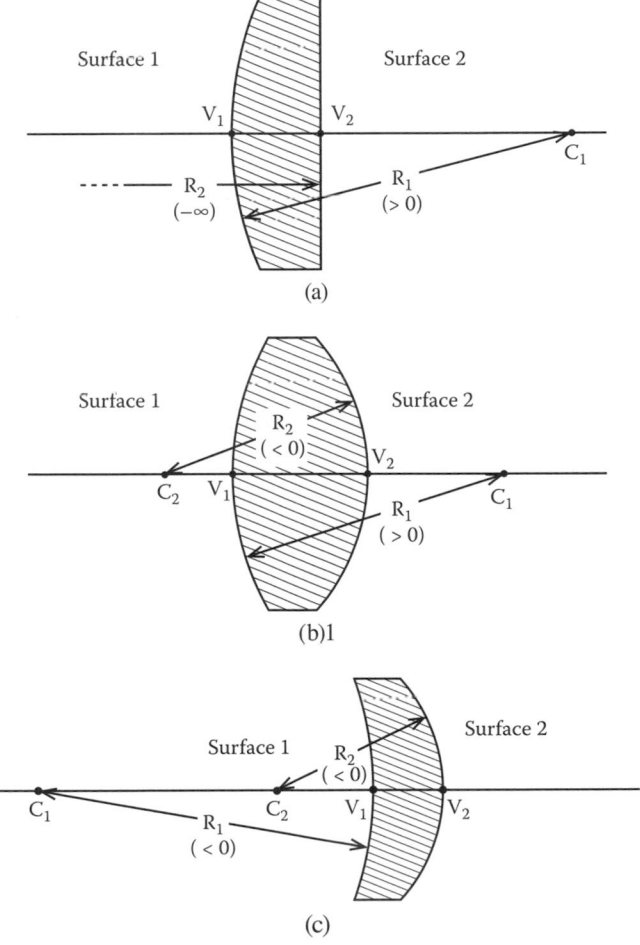

**FIGURE 6.3**  Sign convention for radii for various types of refractive lenses: (a) plano-convex, (b) double-convex (or bi-convex), and (c) concave-convex. When the radial center (C) is to the right of vertex (V), then the radius (R) is positive.

**TABLE 6.2**

**Transfer Matrix Used for Matrix Calculation Method**

|  | Transfer Matrix | Pictorial Representation |
|---|---|---|

Refraction from spherical surface

$$\begin{bmatrix} 1 & -\left(\dfrac{n'-n}{R_1}\right) \\ 0 & 1 \end{bmatrix}$$

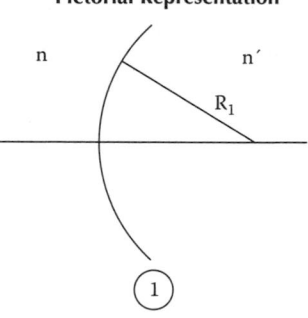

Propagation

$$\begin{bmatrix} 1 & 0 \\ \dfrac{D_{12}}{n_{12}} & 1 \end{bmatrix}$$

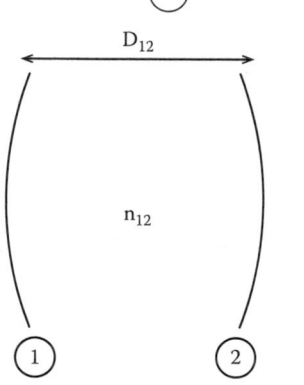

Thick lens

$$\begin{bmatrix} 1-\left(\dfrac{P_2 D_{12}}{n_T}\right) & -P_1 - P_2 + \left(\dfrac{P_1 P_2 D_{12}}{n_T}\right) \\ \dfrac{D_{12}}{n_T} & 1-\left(\dfrac{P_1 D_{12}}{n_T}\right) \end{bmatrix}$$

$$P_1 = \frac{n_T - 1}{R_1}$$

$$P_2 = \frac{n_T - 1}{R_2}$$

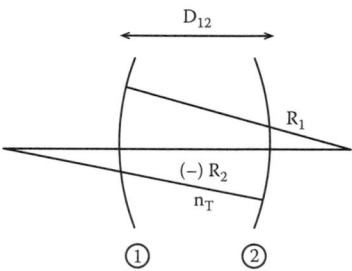

Thin lens
$(D_{12} \to 0)$

$$\begin{bmatrix} 1 & -P_{thin} \\ 0 & 1 \end{bmatrix}$$

$$P_{thin} = \frac{1}{f} = (n_l - 1)\left(\frac{1}{R_1} - \frac{1}{R_2}\right)$$

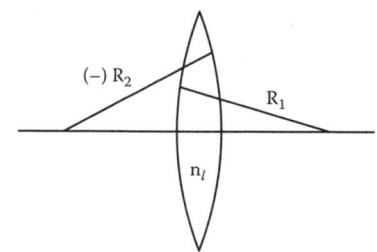

**TABLE 6.3**

**Example of Effective Focal Length and Principal Plane Calculations for a Double Convex Lens**

| Input Parameters | | | Calculated Values | | |
|---|---|---|---|---|---|
| Parameter | Value | Units | Parameter | Value | Units |
| $R_1$ | 9 | mm | $f$ | 11.2 | mm |
| $R_2$ | –12 | mm | $V_1 H_1$ | 1.55 | mm |
| $d$ | 5 | mm | $V_2 H_2$ | –2.07 | mm |
| $n$ | 1.5 | | | | |

## 6.9   BEAM SHAPING FROM LASER DIODE

To convert elliptical output of a laser diode (LD) to a circular beam, two cylindrical lenses can be used as shown in Figure 6.4. Light emission from many laser diodes is elliptical, and it is specified by the emission angles $\theta_A \times \theta_B$. In addition, light output from a laser diode chip is expanding. To convert the LD output to a collimated circular beam, two cylindrical lenses of two different focal lengths are used. The choice of lens focal lengths depend on the emission angle, namely,

$$\theta_A/\theta_B = f_A/f_B \tag{6.10}$$

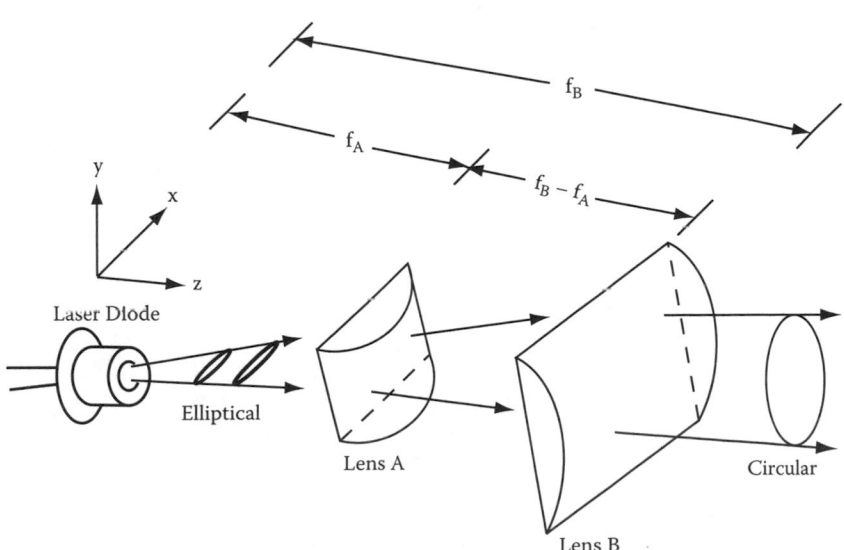

**FIGURE 6.4**   Use of cylindrical lenses to convert elliptical output of a laser diode to a circular beam.

where $f_A$ and $f_B$ refer to the focal length of the fist and second lenses. Lens A is placed at a distance $f_A$ from LD, whereas Lens B is placed at a distance $f_B$ from LD, therefore the distance between the two lenses is $f_B - f_A$. Often plano-convex lenses are used for this purpose, and they cannot be approximated by thin lens formulation. When using plano-convex lenses, the planar surface of the lenses should be facing the LD to minimize aberrations. When using thick lenses, the distances between the lenses are calculated from their back focal lengths, namely, the distance should be $BFL_B - BFL_A$.

Many of the optical component vendors offer matching collimation optics for the particular LD that they carry.

## REFERENCES

1. Hecht, E. 2001. *Optics*. 4th ed. Reading, MA: Addison-Wesley.
2. Kingslake, R., and R. B. Johnson. 2009. *Lens Design Fundamentals*. 2nd ed. Orlando, FL: Academic Press.
3. Klein, M. V., and T. E. Furtak. 1986. *Optics*. 2nd ed. New York: John Wiley & Sons.
4. Mahajan, V. N. 1998. *Optical Imaging and Aberrations, Part I: Ray Geometrical Optics*. Washington: SPIE Press.

# 7 Imaging Systems

The purpose of an imaging system is to collect and focus light scattered from an object thus forming an image of the object, which is then detected either by the human eye, by a photosensitive film, or by a detector array. Imaging systems often magnify or demagnify an object. An example of an imaging system is a camera (one that uses a lens to focus, or a pinhole camera as shown in Figure 7.1) that images an object onto film or a detector array, a telescope that makes far away objects appear nearby, or a microscope that magnifies small objects. Imaging systems utilize a number of refractive (e.g., lenses), reflective (e.g. mirrors), and diffractive components. Examples of various optical components used in imaging systems are shown in Table 7.1.

## 7.1 OPTICAL RESOLUTION

One of the figures of merit to qualify an imaging system is the resolution. Resolution is the measure of how sharply an image can be generated. It is a measure of the spatial frequency response of the imaging system.

The fundamental limiting parameter of the resolution of a conventional optical imaging system is the optical wavelength. In most cases however, resolution is further limited by the numerical aperture (NA) of the system, given by

$$NA = n_o \sin(\theta_{max})$$  (7.1)

where $n_o$ is the refractive index of the medium, and $\theta_{max}$ is largest half angle of light rays that can pass through the optical system. For example, in a single lens imaging system without any iris, $\theta_{max}$ is the angle determined from object distance, $s_1$, and the lens radius, $r$, namely, $\theta_{max} = \tan^{-1}(r/s_1)$.

Numerical aperture is a result of diffraction theory. Finer features diffract light at higher angles than coarse features. Therefore the ability of the system to resolve this feature is dependent on the largest collection angle of the optical system.

Imaging systems are often specified by their f-number ($f/\#$), which is related to the numerical aperture by

$$f/\# = \frac{1}{2 \cdot NA}$$  (7.2)

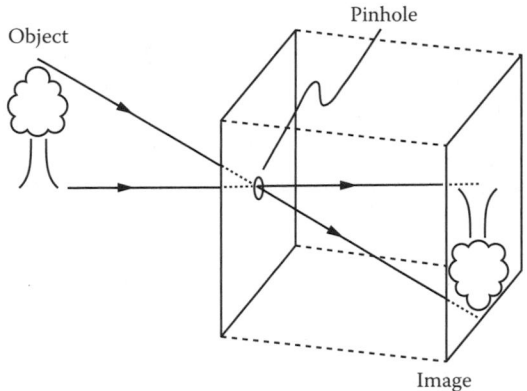

**FIGURE 7.1**   Pinhole camera.

**TABLE 7.1**
**Various Types of Optical Components Used in Imaging Systems**

| Component | What Is It or What Does It Do? | Example Applications |
|---|---|---|
| Spherical lens | Refracts light (changes optical path), thus enabling focusing and defocusing. | Cameras telescopes, microscopes |
| Cylindrical lens | Change optical path in one direction, for example, in x-direction in an x-y-z, where z is the propagation direction. | Converting elliptical light output from a laser diode to circular beam |
| Ball lenses | Lens made of glass ball with very tight focus. | Fiber-to-waveguide coupling |
| Gradient index lenses | A rod-shaped optics, where the refractive index changes from center of the rod outward. It can be used to focus light, with the advantage of not requiring curved surfaces. | Contact imaging sensor (array), fiber collimator, fiber-to-waveguide coupling |
| Curved mirrors | Spherical or curved shaped mirrored surfaces that are used for changing light path, such as for focusing light. | Mirrors of a reflective telescope; broadband microscope objectives |

A good test of optical resolution is modulation transfer function (MTF) [1,2], which can be measured by spatial modulation signal, such as a periodic line pattern:

$$MTF = \frac{I_{max} - I_{min}}{I_{max} + I_{min}} \qquad (7.3)$$

where $I_{min}$ and $I_{max}$ refer to minimum and maximum of the spatial modulation signal, respectively.

MTF measures of the spatial frequency response of the imaging system. Typically a test pattern is used to measure MTF, such as the 1951 USAF target or any appropriate resolution target. Choice of correct resolution target is key to testing the characterizing the MTF of the optical system. An example of the MTF test procedure is to capture an image of the resolution test pattern with the optical imaging system under test, and calculate MTF versus spatial frequency (typically given in line pairs per millimeter [lp/mm] or dots per inch). Some of the fine lines in the resolution test pattern will not be resolved, namely, multiple thin lines will appear to be fused together, indicating the resolution limit of the optical system. This will be apparent in the MTF graph, where MTF reaches zero beyond spatial frequencies where the system could resolve lines.

An example of MTF measurement, shown in Figure 7.2. Figure 7.2a and Figure 7.2b, illustrates how an image can deteriorate due to defocusing. Figure 7.2a is the original image with a resolution test target of period 2 mm to 0.25 mm, or the corresponding spatial resolution ranging from 0.5 to 4 lp/mm. Figure 7.2b shows an image that is distorted due to intentional defocusing of the camera lens. The resulting MTF measurements are shown in Figure 7.2c,d. Figure 7.2c shows amplitude of the resolution test pattern for focused and defocused images. From this data, MTF is calculated and plotted in Figure 7.2d. MTF drops much more rapidly for the defocused image than the focused one.

The *Resolution* MATLAB® lines per millimeter illustrates the effect of image blur due to limited MTF (given in lines/mm in the simulation graphic user interface).

When measuring and calculating MTF, care should be taken to make a note of the base level of the signal. Many optical systems that use electronic imaging, such as photodetector arrays, may have high background voltage, which if not biased properly can appear as a background signal ($I_{min}$). This background is due to electronics. Therefore, if the goal is to measure true optical MTF, test conditions should address this, as $I_{min}$ in Equation (7.3) will be affected both by optics and electronics. A technique to overcome this is to measure the dark signal ($I_{dark}$), signal at maximum (e.g., using a white paper, $I_{white}$), and compare this to the spatially modulated signal:

$$MTF = \frac{I_{max} - I_{min}}{I_{white} + I_{dark}} \qquad (7.4)$$

Another factor that affects resolution is aberration in the optical systems. Aberrations [3,4] in the optical system cause blurring or distorting of the image, and therefore reducing resolution. Aberrations are discussed in Section 7.5 and Section 7.6

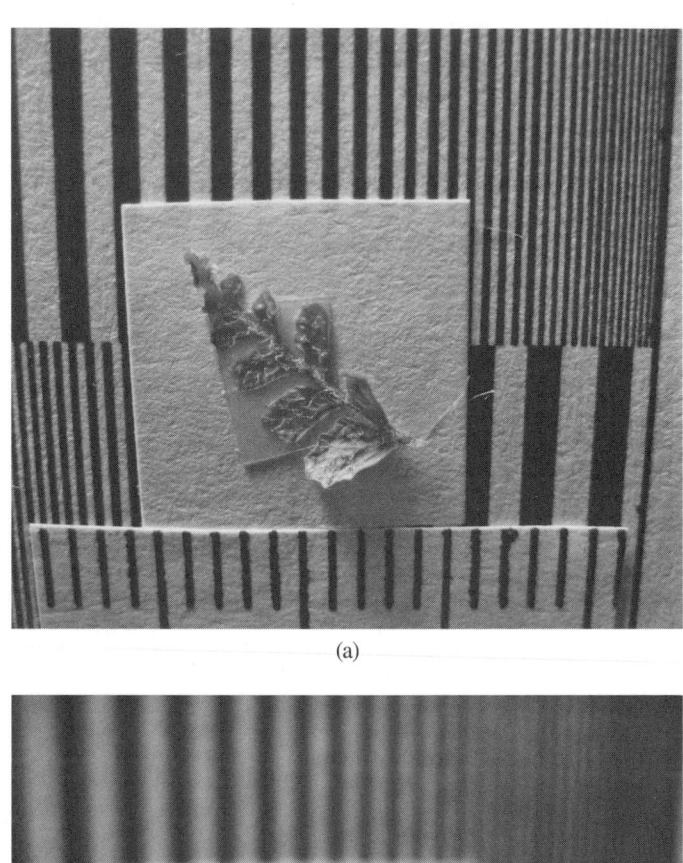

(a)

(b)

**FIGURE 7.2** Use of resolution test pattern to determine image quality. (a) Focused image of a resolution test pattern and an object. Bottom scale: 1 mm per graduation. Image taken with a digital camera. (b) Same target capture while the camera is out-of-focus. (c) Line scan of the top portion of the image. Amplitude modulation of the defocused image clearly drops off much faster than the focused image. (d) Modulation transfer function calculated from the minimum and maximum values of amplitude modulated signal shown in (c). *(continued)*

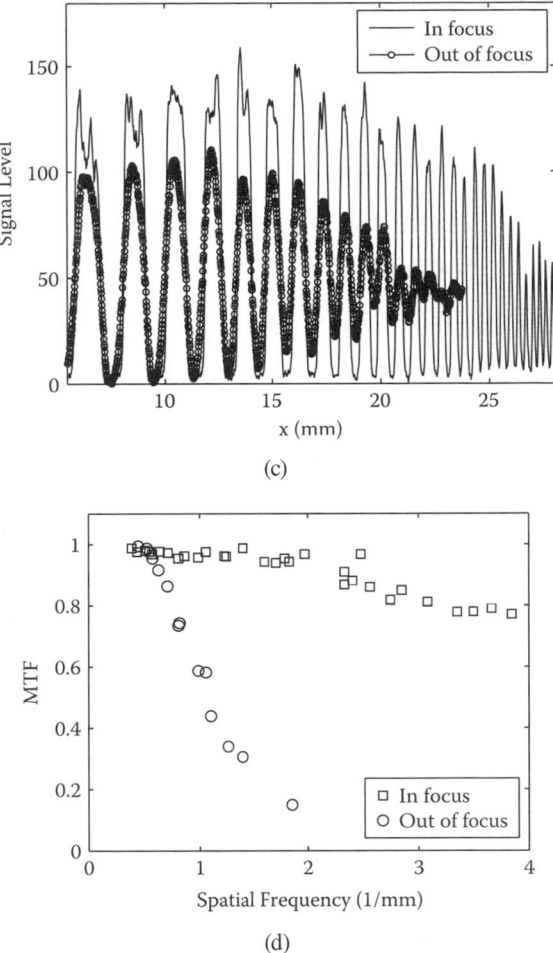

**FIGURE 7.2**   *(continued)* Use of resolution test pattern to determine image quality. (a) Focused image of a resolution test pattern and an object. Bottom scale: 1 mm per graduation. Image taken with a digital camera. (b) Same target capture while the camera is out-of-focus. (c) Line scan of the top portion of the image. Amplitude modulation of the defocused image clearly drops off much faster than the focused image. (d) Modulation transfer function calculated from the minimum and maximum values of amplitude modulated signal shown in (c).

## 7.2   TWO-DIMENSIONAL IMAGING SYSTEMS

Examples of 2-dimensional (2-D) imaging systems are cameras, microscopes, binoculars, and telescopes. These systems consist of imaging optics, refractive lenses, mirrors, diffractive optics, or a combination of several of these technologies. Imaging systems that rely on a human observer, such as a nondigital telescope or microscope, have to take into account the optics of the eye of the human observer. On the other hand, many modern-day instruments utilize photographic film (modern in the sense of "geologic time") or a 2-D detector array. The optics image the object onto the detector array, which then converts the light intensity to an electrical signal, which is then displayed on a screen or stored electronically.

Conventionally, imaging systems relied on spherical surfaces such as mirrors and lenses, because it is a convenient way of grating and producing high-quality optics. Modern-day optics also utilize a variety of optics, such as lenslet arrays, lenses that are integrated with the detector array, lenses that are molded or fabricated with precision mechanics to produce aspheric surfaces. Large telescopes also utilize polygon mirrors, where each section of the mirror can be shifted in real-time to produce sharp images and correct for changes in light path due to atmospheric turbulence.

## 7.3 ONE-DIMENSIONAL IMAGING SYSTEMS; LINE SCAN SENSORS

Line scan sensors convert light intensity into a line image in one dimension. One example of an application of a line scanner is a contact image sensor (CIS). CIS is a sensor used in scanners, copiers, and fax machines to image a document and convert it into a digital format. It utilizes a line scan sensor and imaging optics that image a linear section of the document. The signal for each line scan is stored in digital format, and then as the document moves, consecutive line scans are combined to produce a digital image, as shown in Figure 7.3.

The imaging optics in CIS is an array of spherical lenses. Each lens creates an image of portion of the document onto the sensor, and an array detector converts that

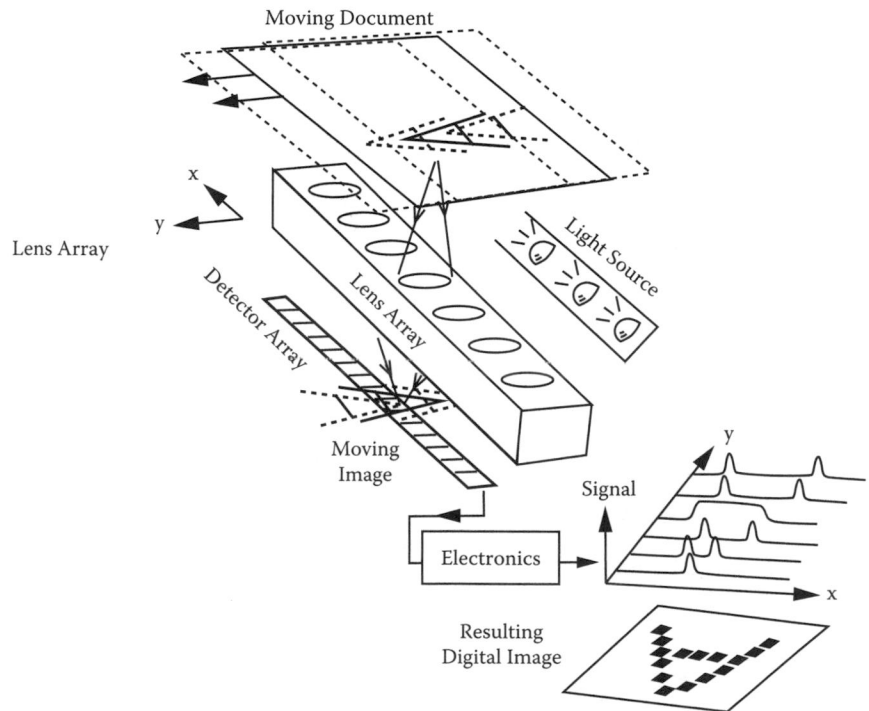

**FIGURE 7.3**   Illustration of image generation using a linear array detector.

portion of the image to a linear signal, which is then digitized and stored. The reason for using a linear array of lenses compared to a single large lens is to fit the sensor in a compact area while producing a quality image.

One of the components found in many of the commercially available CIS is gradient index (GRIN) lens array [5]. A GRIN lens acts similar to a spherical lens, but achieves imaging without having the need to have a curved surface. Each lens in the array is a glass rod, with the refractive index varying in the radial direction. The refractive index is highest in the center and decreases gradually toward the outer region of the lens. By controlling the refractive index profile and the length of the rod, the GRIN lens can be used to produce a 1:1 image. Variations of GRIN lenses can also be used for other applications, such as for collimating light coming out of a fiber optic [5].

## 7.4   STOPS

All optical systems have a limited aperture due to the limited diameter of the optical components. Some optical systems, such as cameras, also incorporate additional apertures. If the aperture is used to limit the amount light from reaching the detector, while allowing the full field of view to be imaged, it is called *aperture stop* (A.S.). A.S. is used in cameras to adjust to various degrees of brightness. An optical system with a focal length, $f$, and diameter, $D$, is referred to by its *relative aperture*, or *f-number*:

$$f/\# = \frac{f}{D} \tag{7.5}$$

The irradiance at the image plane (in W/m$^2$, also referred to as radiant flux density) varies by the square of the inverse of the $f/\#$, namely, by $(D/f)^2$.

Another type of element is called *field stop*, which limits the field of view of the optical system. The position of the aperture in the optical system determines if it is a field stop or an aperture stop. Detailed discussion on stops and pupil can be found in References [2, 6-8].

## 7.5   MONOCHROMATIC ABERRATIONS

There are number factors that affect the performance of the quality of optical systems. For example, rays that are not all focused to one spot, if the image is distorted, or if the image focal spot varies by wavelength. These phenomena are called aberrations. The subject of aberrations is an advanced subject of optics. For completeness, we will only glance at this topic, but for detailed understanding the reader is referred to other optics text that describe aberrations in great detail [3,4,6–10].

Primary monochromatic aberrations are

- Spherical
- Coma
- Astigmatism
- Distortion
- Field curvature

The description of these aberrations, their corresponding figures (Figures 7.4 to 7.8), and suggested corrective actions are listed in Table 7.2.

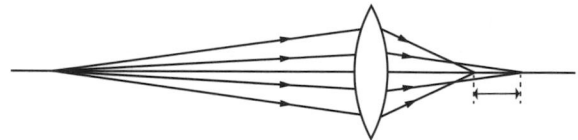

**FIGURE 7.4**   Spherical aberration.

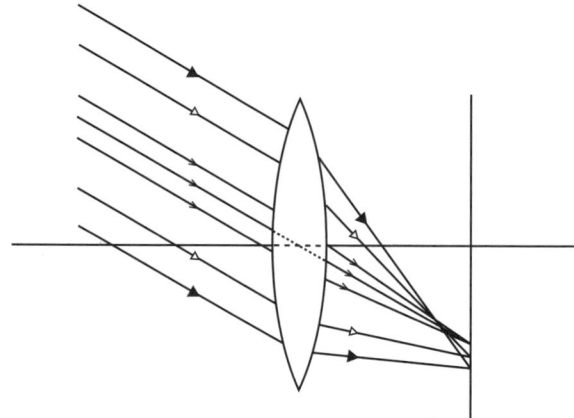

**FIGURE 7.5**   Coma.

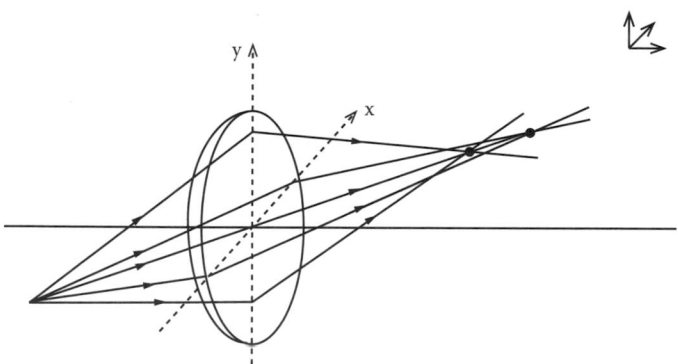

**FIGURE 7.6**   Astigmatism.

(a)

**FIGURE 7.7**   Demonstration of distortion. (a) Original image. (b) Barrel distortion. (c) Pincushion distortion. *(continued)*

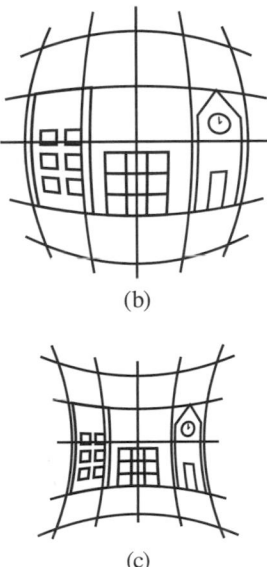

(b)

(c)

**FIGURE 7.7** *(continued)* Demonstration of distortion. (a) Original image. (b) Barrel distortion. (c) Pincushion distortion.

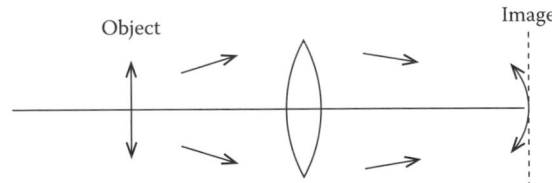

Object

Image

**FIGURE 7.8** Curvature of field.

## 7.6 CHROMATIC ABERRATIONS

Chromatic aberrations are aberrations that are due to the fact that the refractive index of an optical element are wavelength dependent (dispersion), and therefore the focal point of a single lens is different at different wavelengths, as shown in Figure 7.9. Chromatic aberrations are particularly important factors for imaging using broad spectral illumination, such as the visible spectrum. A key parameter used for characterizing glass materials is the Abbe number, also called V-number or dispersive index, given by

$$V_d = \frac{n_d - 1}{n_F - n_C} \tag{7.6}$$

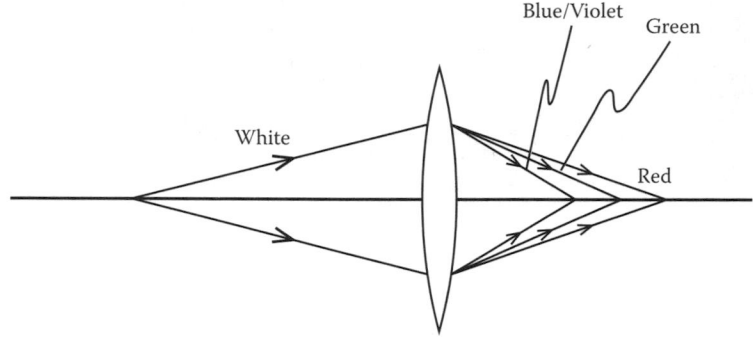

**FIGURE 7.9**   Chromatic aberration.

where $n_d$ refers to refractive index at yellow light (at wavelength 588 nm He line), $n_F$ and $n_C$ refer to refractive indices at blue light (at wavelength 486 nm H line) and red light (at wavelength 656 nm H line), respectively. Chromatic aberrations can be mini-mized by using a pair of positive and negative lenses of different materials [7], given by

$$f_{1d}V_{1d} + f_{2d}V_{2d} = 0 \tag{7.7}$$

where $f_{1d}$ and $V_{1d}$ refer to the focal length and $V_d$ of the first lens, and $f_{2d}$ and $V_{2d}$ refer to the focal length and $V_d$ of the second lens.

In addition to aberrations associated with conventional lenses, new imaging devices produce rather unconventional distortions, such as shown in Figure 7.10.

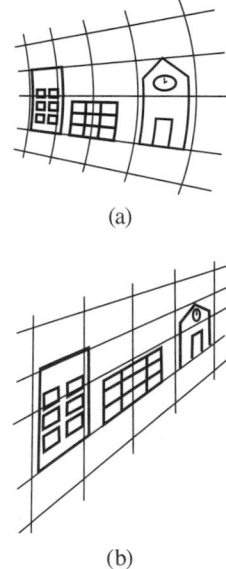

(a)

(b)

**FIGURE 7.10**   Other types of distortion observed in digital cameras.  Original image is shown in Figure 7.7.

Devices such as cell phones cameras often use a combination of optics and image processing to create images. Some of the modern devices may be using lenslet arrays rather than a single lens, which can create sharp images in conditions where cameras with conventional optics may not be able. Some of these "aberrations" are either due to optics, to the system, or a combination of both, and many of them are corrected by image processing.

Aberration correction involves optical design and is described in detail in various textbooks such as by Kingslake [3]. Many of the aberrations are due to the fact that lenses are generally spherical. A nonspherical lens (called aspheres) is another type of lens that can minimize certain types of aberrations, such as spherical aberrations.

Table 7.2 lists some of the aberrations and "quick fixes" if such aberrations are encountered. Complete aberration correction requires detailed optical design, which is described in various optical engineering text [3,6–8].

**TABLE 7.2**
**Primary Aberration Descriptions and Some Corrective Actions**

| Aberration | Brief Description | Some Corrective Action (Quick Fixes) |
|---|---|---|
| • Spherical aberration (Figure 7.4) | • Paraxial rays (rays that at small angle with respect to the optical axis) and nonparaxial rays focus at different distances. | • Using aspheric lenses <br> • Using multielement system <br> • Increase focal length for a single-lens imaging |
| • Coma (Figure 7.5) | • Rays that travel at extremities of the imaging system arrive at different locations than the principal ray (rays that pass through the center of the lens, also known as principal point). Coma is generally associated with off-axis rays. | • Minimize off-axis rays <br> • Increase focal length for single-lens imaging or use a multielement system instead |
| • Astigmatism (Figure 7.6) | • If the object is away from the optical axis in both x- and y-directions, then a point source is focused to a line that varies direction as light propagates in the z-direction. | • If possible, minimize off-axis rays <br> • Asymmetric lenses could minimize this aberration <br> • Use cylindrical and spherical lenses |
| • Distortion (Figure 7.7) | • Distortion is aberration where the image looks distorted, primarily because image magnification is nonuniform across the image plane. Barrel distortion and pincushion distortion are two types of distortion. | • Minimize use of small radii lenses <br> • Use multielement lens system instead of a single element focusing system. <br> • For digital imaging, may be able to compensate using image processing and digital image correction. |

(continued)

**TABLE 7.2 (CONTINUED)**
**Primary Aberration Descriptions and Some Corrective Actions**

| Aberration | Brief Description | Some Corrective Action (Quick Fixes) |
|---|---|---|
| • Field curvature (Figure 7.8) | • Field curvature occurs when the object is imaged on a curved plane. Field curvature in many biological imaging systems is compensated by the curved image plane (such as the retina in the human eye). However, this is not a luxury that most artificial imaging systems have (such as cameras) and needs to be corrected with optical elements. | • Minimize use of small radii lenses<br>• Use multielement lens system instead of a single element focusing system |
| • Chromatic aberration | • Focal spot is different for different wavelengths, due to dispersive properties of lenses (because refractive index of glass is wavelength dependent). | • Combine positive and negative lenses of different dispersive power ($V_d$), such as crown and flint glass (achromatic doublet, or achromat) |

## 7.7  VARIOUS TYPES OF ILLUMINATION

Illumination plays a key role in imaging systems. Various types of illumination are described next.

### 7.7.1  COHERENT ILLUMINATION

An example of coherent illumination is using a laser light source, such as used for interferometry. Coherent illumination refers to light with phase that varies in unison. When phase of a light wave varies in identical fashion at two points in *space*, the illumination is said to be *spatially coherent*. When phase of a light wave varies in identical fashion at two points in *time* (i.e., at two separate points along the direction light propagation), then it is said to be *temporally coherent*. One advantage of using a laser is that it can be focused to a tight spot, enabling concentration of high optical power onto a small spot. For interferometry, coherent illumination makes is easier for aligning optics and is forgiving to variations in differences in optical path length. Namely, interference fringes can be obtained as long as the reference and the object light paths of an interferometer are within the coherence length of the laser. One of the drawbacks of laser illumination is speckle noise. This can be a hindrance to imaging systems, where speckle noise can deteriorate image quality. Some of this noise can be compensated by using time averaging and moving diffusers in the optical setup, as described in Chapter 4 (Section 4.6.3).

### 7.7.2 Incoherent or Partially Coherent Illumination

Incoherent or partially coherent illumination, such as illumination from LED and white light sources, can create desirable images. For certain applications, such as intereferometric microscopy, this can be a challenge, as the reference and object beam of the interferometer has to be kept within the coherence length, which could be in the micrometer range. Some methods to overcome this challenge include using narrow band optical filters following the partially coherent source to further increase the coherence length (the narrower the filter, the longer the coherence length). Another technique is to use an interferometer objective lens, where the reference and object beams are equal when the object is in focus.

### 7.7.3 Point Source and Diffuse Illumination and Multiangle Illumination

Illumination has to be designed according to the imaging requirements of the system. For example, if an edge enhanced image is desired, high-angle illumination and point sources are utilized. (Point source is a light source that appears to be "infinitely" small at the point of emission. Examples of point source illumination are a halogen lamp placed at some distance from the object and direct sunlight). If uniform image is desired with minimum variation in image brightness due to object topography, then diffuse illumination and illumination from multiple angles is used. Figure 7.11 illustrates the effect of illumination on the appearance of a wrinkled paper. Figure 7.11a shows the wrinkled paper illuminated with a light source from

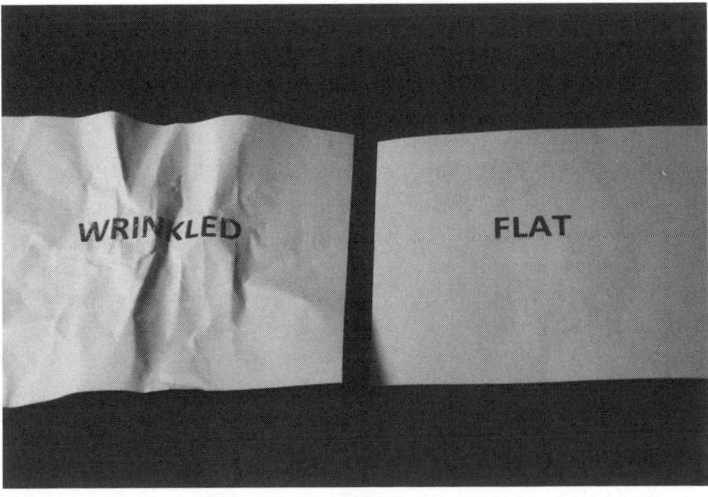

(a)

**FIGURE 7.11** Effect of illumination on the appearance of wrinkled and flat paper. (a) Papers illuminated with a light source from *left* side, and wrinkles are clearly visible. (b) Papers illuminated with a light source from *right* side, and wrinkles are clearly visible. (c) Papers illuminated from *left* and *right side*, and wrinkle noise is minimized. *(continued)*

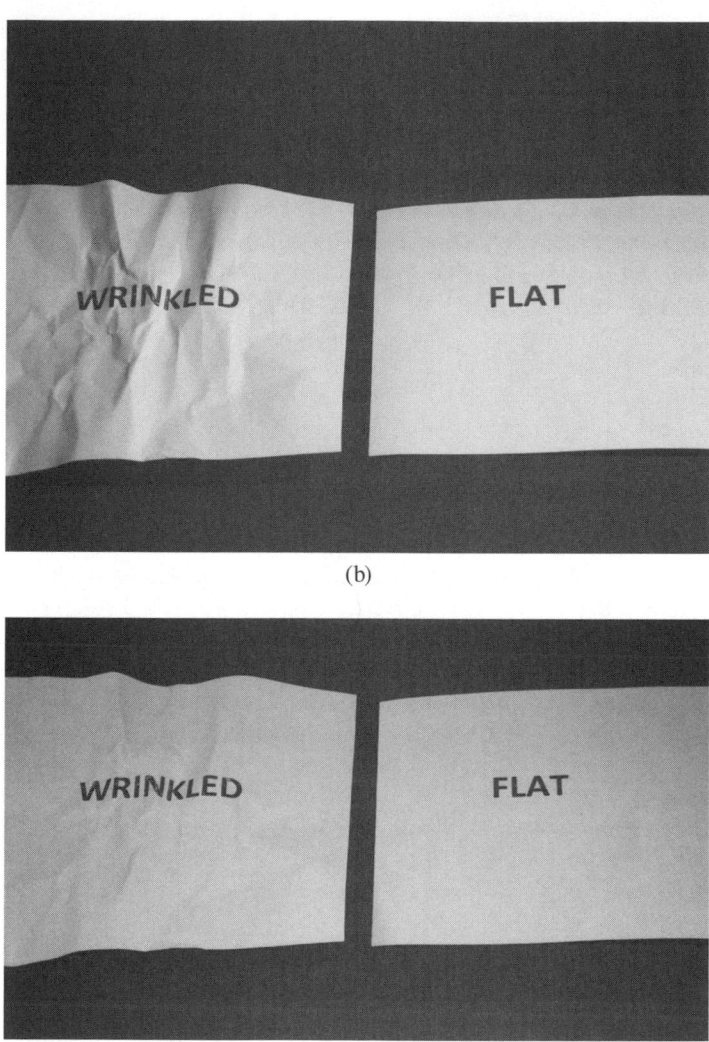

(b)

(c)

**FIGURE 7.11** *(continued)* Effect of illumination on the appearance of wrinkled and flat paper. (a) Papers illuminated with a light source from *left* side, and wrinkles are clearly visible. (b) Papers illuminated with a light source from *right* side, and wrinkles are clearly visible. (c) Papers illuminated from *left* and *right side*, and wrinkle noise is minimized.

the left side. As apparent, wrinkles are highly visible in the image. Figure 7.11b shows the wrinkled paper illuminated with a light source from the right side. Wrinkles are also visible in this image. When both left and right side light sources are turned on and paper is illuminated from both sides, wrinkle noise is minimized, as shown in Figure 7.11c. In fact, if the world was illuminated by diffuse sources from all directions, one would not have to iron their shirt, as wrinkles would not be visible. The drawback, however, is that many objects would appear dull, as contrast created by shadows gives character to the surrounding scenery.

## REFERENCES

1. Goodman, J. W. 2004. *Introduction to Fourier Optics*. 3rd ed. Greenwood Village, CO: Roberts & Company.
2. Mahajan, V. N. 2011. *Optical Imaging and Aberrations, Part II: Wave Diffraction Optics*. 2nd ed. Washington: SPIE Press.
3. Kingslake, R. and R. B. Johnson. 2009. *Lens Design Fundamentals*. 2nd ed. Orlando, FL: Academic Press.
4. Born, M., and E. Wolf. 1999. *Principles of Optics: Electromagnetic Theory of Propagation, Interference and Diffraction of Light*. 7th ed. New York: Pergamon Press.
5. Yeh, P. 2005. *Optical Waves in Layered Media*. 2nd ed. New York: Wiley Interscience.
6. Jenkins, F. A., and H. E. White. 1976. *Fundamentals of Optics*. 4th ed. New York: McGraw-Hill Book Company.
7. Hecht, E. 2001. *Optics*. 4th ed. Reading, MA: Addison-Wesley.
8. Klein, M. V., and T. E. Furtak. 1986. *Optics*. 2nd ed. New York: John Wiley & Sons.
9. Mahajan, V. N. 1998. *Optical Imaging and Aberrations, Part I: Ray Geometrical Optics*. Bellingham, WA: SPIE Press.
10. Malacara, D. 2007. *Optical Shop Testing*. 3rd ed. New York: Wiley Interscience.

# 8 Guiding Lightwaves

## 8.1 LIGHT GUIDING AND TOTAL INTERNAL REFLECTION

As discussed in Chapter 4, the refraction angle is given by Snell's law [1]:

$$n_1 \sin \theta_1 = n_2 \sin \theta_2 \tag{8.1}$$

When light is traveling in a material such as glass or water, which has higher refractive index than its surroundings, such as air, then beyond a certain angle of incidence called critical angle ($q_c$), light completely reflects back. Namely, when the refracted angle in the second medium (e.g., air) is 90 degrees, then the incident angle in the first medium is at critical angle, which is the angle beyond which light cannot escape.

$$\sin \theta_c = \frac{n_1}{n_2} \tag{8.2}$$

For example, a scuba driver looking up from under water can see what is above water, such as a seagull or pelican (see Figure 8.1a). When the diver moves up close to the surface trying to look outside at a shallow angle, the surface looks like a mirror, and the diver cannot see the seagull (see Figure 8.1b). This is because the viewing angle is now larger than the critical angle. The refractive index for water (in the visible spectrum) is 1.33. Therefore the critical angle for water is $\sin^{-1}(1/1.33) = 48.8°$.

The same phenomenon occurs in glass, where light traveling from glass to air cannot escape if the angle of incidence is larger than the critical angle. (For a glass with a refractive index of 1.5, the critical angle is 41.8°). This phenomenon is called total internal reflection (see Figure 8.2).

If both sides of the glass are surrounded by air, such as for a glass slide, light rays with incident angle larger than the critical angle are trapped in the slab, as shown in Figure 8.3. This phenomenon is call light guiding.

If the light guide bends slowly (adiabatic), such that the incident angle of the light rays is larger than the critical angle, then light will continue to be guided, as shown in Figure 8.4.

On the other hand, if the light guide is bent sharply, such as at a small radius shown in Figure 8.5, then light rays will escape when the angle of incidence is less than the critical angle.

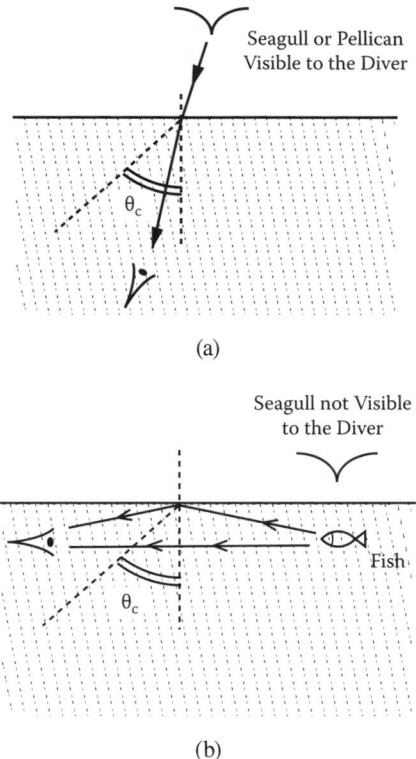

**FIGURE 8.1** Underwater total internal reflection. (a) Diver can see the seagull because the viewing angle is less than the critical angle. (b) Diver can see the fish and the reflection from the water surface due to total internal reflection but cannot see the seagull because the viewing angle is larger than the critical angle. At this position the surface of the water looks like a mirror to the diver.

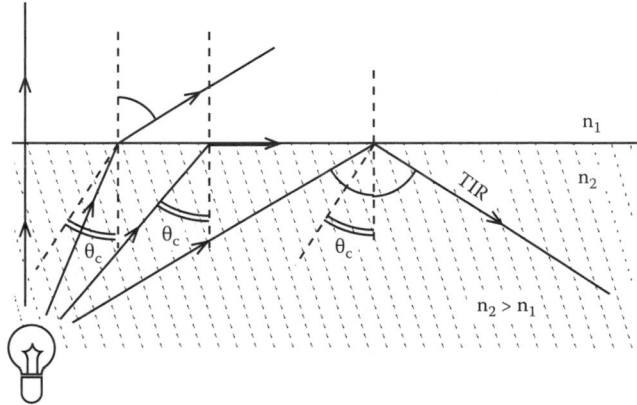

**FIGURE 8.2** Light rays propagating from high refractive index ($n_2$) medium (e.g., glass; $n_2 =$ 1.5) to low refractive index ($n_1$) medium (e.g., air; $n_1 = 1$). When the incident angle is equal to the critical angle ($\theta_c$), light is trapped at the interface of the two media. When the incident angle is less than the critical angle, total internal reflection (TIR) occurs, and no light escapes to the medium with lower refractive index.

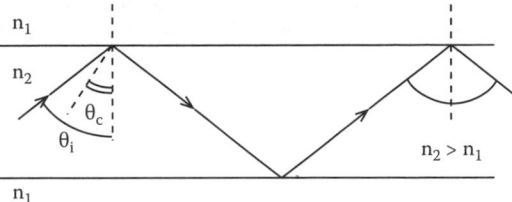

**FIGURE 8.3** Light trapped (guided) via total internal reflection in a medium of high refractive index ($n_2$), surrounded by lower refractive index ($n_1$) media. This phenomenon occurs when the angle of incident ($\theta_i$) is larger than the critical angle ($\theta_c$).

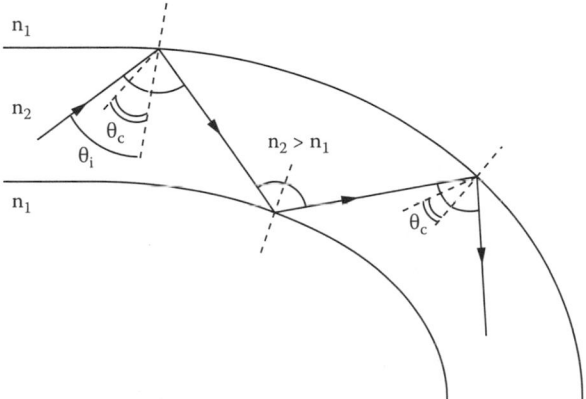

**FIGURE 8.4** Light trapped in a slowly (adiabatic) bending light guide.

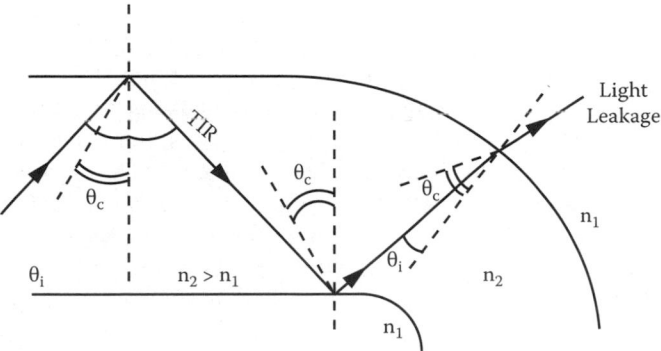

**FIGURE 8.5** Light trapped in a light guide can escape due to sharp bends because angle of incident ($\theta_i$) is smaller than the critical angle ($\theta_c$).

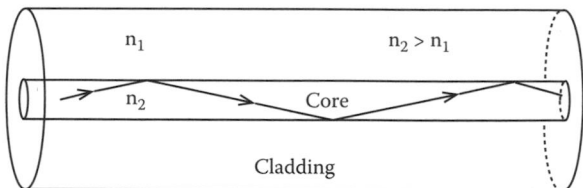

**FIGURE 8.6**   A fiber optic light guide where the core has higher refractive index ($n_2$) than the cladding ($n_1$). Light is guided inside the core.

## 8.2   FIBER OPTICS

With the advent of fiber optics, many of the communication lines conventionally made of copper wires were replaced with fiber optics because of the advantage optical fibers have of larger bandwidth, smaller cable, and lower loss than conventional copper wires [2–4].

Fiber optics are also based on the principle of total internal reflection. If a single layer of glass or plastic fiber is surrounded by air, light will be trapped in the fiber via total internal reflection. However, if the interface of a single layer of fiber gets perturbed, for example, if the face gets dirty or gets scratched, then light will escape, causing loss of optical signal. A solution to this is to have a layered structure, where the core of the fiber has higher refractive index than the surrounding (called cladding), as shown in Figure 8.6. In this case light is trapped in the core, and very little or almost no light reaches the outer layer of the cladding. Therefore a fiber with a core–cladding structure is less likely to incur any losses due to perturbations at the cladding–air interface.

Optical fibers come in various formats. For long-distance communication, a common fiber is a single-mode fiber, where only one mode of light is trapped. The core and cladding layers are approximately 9 microns and 125 microns in diameter, respectively. Typically these fibers have very low loss for telecommunication wavelengths of 1.3 microns and 1.55 microns. For short distance communication, and for other applications, such as for sensing and for lighting, multimode fibers are also used. Examples of multimode fibers are 65 micron and 900 micron core fibers. A relatively new class of fibers called photonic crystal fibers achieve light trapping by incorporating small periodically structured holes that run parallel inside the fiber, instead of using core-cladding structures.

## 8.3   PLANAR WAVEGUIDES AND INTEGRATED OPTICS

Light guiding can be achieved in a planar structure, where light can be trapped in planar direction. Light can be trapped in multilayers by surrounding a layer of material with high refractive index (called core layer) with layers of lower refractive index

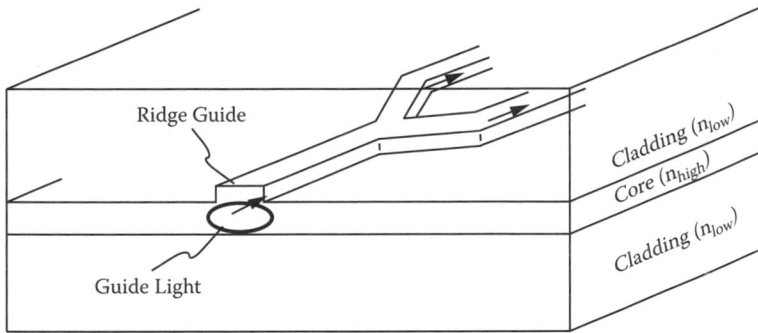

**FIGURE 8.7**    Integrated waveguide structure.

(called cladding layers). This allows confinement of light in the core. Further confinement can be achieved using ridges or strips that can be etched or embedded either in the core or cladding layers. An example of a ridge waveguide structure is shown in Figure 8.7. Under the ridge, the effective refractive index [5,6] is higher than in the area outside the rib, therefore this difference allows light guiding along the ridge.

## 8.4   COUPLING BETWEEN FIBERS AND WAVEGUIDES

A number of approaches are available to couple light between fibers, splitting light from one waveguide to multiple waveguides, and between fibers and waveguides, and from free space optics to and from waveguides. Figures 8.8 to 8.13 show some of these techniques [5–10].

Often when coupling from fibers to waveguide, there is a mismatch in the mode shape. The fiber mode tends to be circular, whereas many planar waveguides and lasers have elliptical mode shape, and often much smaller than the fiber mode. To

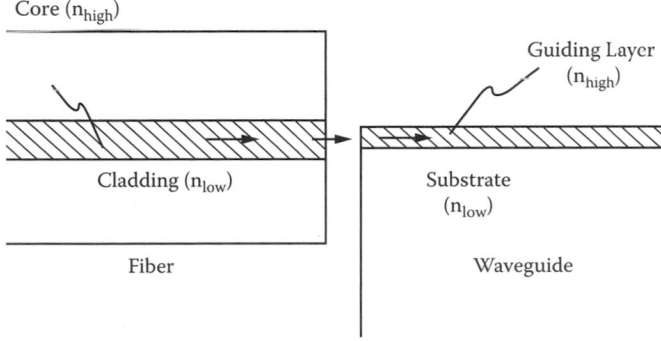

**FIGURE 8.8**    Fiber to waveguide end-fire coupling.

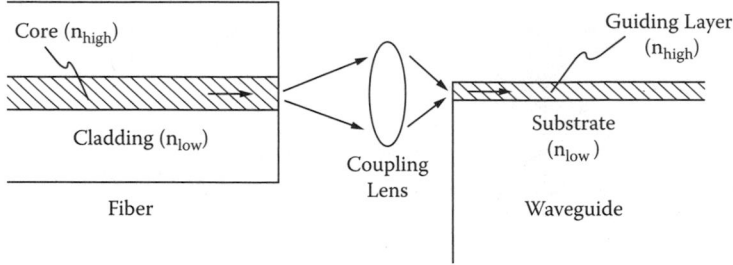

**FIGURE 8.9**  Fiber to waveguide coupling with a lens. The lens could be a refractive lens, a ball lens that is a glass sphere used as a lens, or a gradient index lens, or a multielement lens.

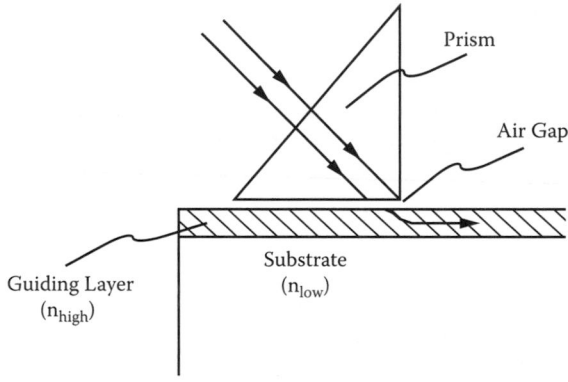

**FIGURE 8.10**  Free space to waveguide prism coupling.

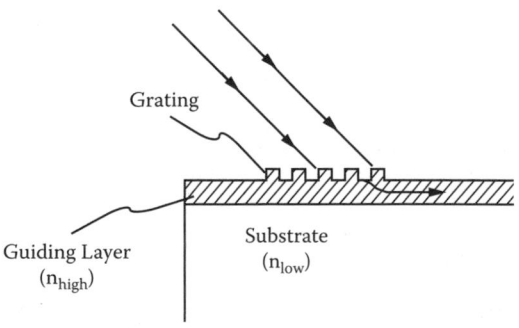

**FIGURE 8.11**  Free space to waveguide grating coupling.

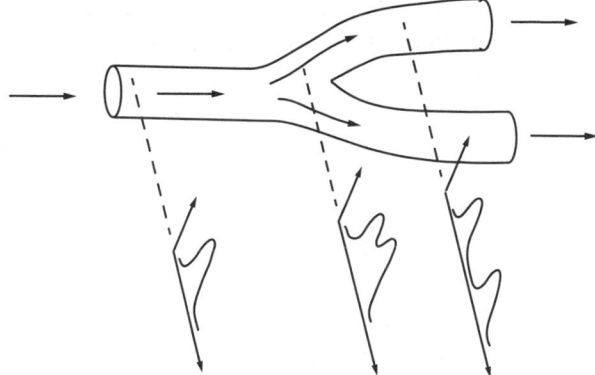

**FIGURE 8.12**   Y-junction coupler.

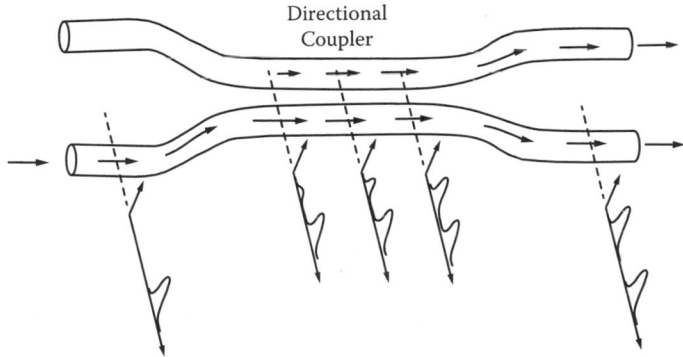

**FIGURE 8.13**   Directional coupler.

minimize coupling loss, various techniques can be used, such as using intermediate non-standard fiber (such as high-numerical aperture) between the waveguide and fiber, using tapered fibers, and taper waveguides.

## 8.5   ACTIVE INTEGRATED OPTICAL DEVICES

Guided wave devices can be made active, such as making optical switches and modulators. Because of their construction with thin waveguide layers, guided wave devices are ideal to incorporate active elements using electro-optic or thermo-optic

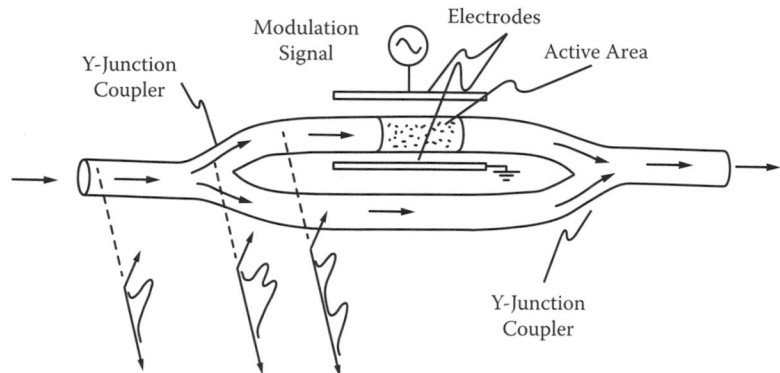

**FIGURE 8.14**    Electro-optic modulator in an active Mach-Zehnder waveguide interferometer configuration.

effects. For example, an electro-optic modulator is a device that incorporates linear electro-optic (EO) material, where voltage applied across an EO layer induces change in refractive index. The change in refractive index is dependent in the electric field amplitude (given in volt per meter [V/m]), and the thinner the material, the larger the electric field, and the larger the induce change in refractive index. If the device is configured in a Mach-Zehnder (MZ) interferometer configuration, as shown in Figure 8.14, and when a modulation voltage is applied across an active EO layer, the change in refractive index results in a change in the intensity of the transmitted light. Therefore an EO MZ device can be used to modulate incoming light with a high-speed radio frequency (RF) source. Examples of such devices are made from $LiNbO_3$ materials as well as EO polymers [9]. Other examples of active waveguide devices are thermo-optic switches.

Passive and active elements are combined in an integrated structure to form photonic integrated circuits used for variety of telecommunication applications [11–13].

## 8.6   SUGGESTED SIMULATIONS

Simulation 8.1: Total Internal Reflection

Using the Refraction MATLAB® Simulation, enter $n_1 = 1.5$ (glass), and $n_2 = 1$ (air). Start with a small angle of incidence, such as 10 degrees, and note the refraction angle. Next, increase the angle and observe that beyond a certain angle the light rays reflect back. This occurs when $\theta_1$ is larger than critical angle, $\sin^{-1}(1/1.5) = 41.8°$.

## 8.7   SUGGESTED EXPERIMENTS

### EXPERIMENT 8.1: GUIDING LIGHT USING A SLAB WAVEGUIDE

Purpose: To familiarize with light guiding and light out-coupling
Related chapter/section: Guiding lightwaves

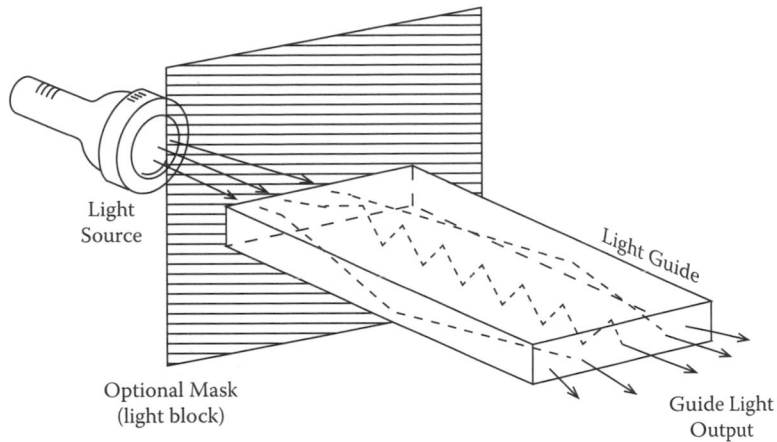

Light
Source

Optional Mask
(light block)

Light Guide

Guide Light
Output

**FIGURE 8.15**   Guiding light using slab waveguide.


Related simulations: N/A
Materials needed:

- Flat rectangular glass slab. Example: 1 in × 3 in × 1 mm microscope slide
- One of the following:
    - Paint, or liquid white out, or
    - Diffuse tape, such as Scotch® Magic™ Tape 810
- Flashlight
- Light block: Foil or black cardboard as shown in Figure 8.15


### EXPERIMENT

1. Light propagating in a waveguide—Set up the experiment shown in Figure 8.15, starting with a clean (unpainted) light guide, such as a slab waveguide. A 1 mm thick microscope slide is a good example. Couple light from the flashlight at one end, and observe guided light output coming out at the other end. Use a block (such as cardboard or foil or black paper) to block the unguided light with a screen, as shown in Figure 8.15. A simple step to accomplish this is to make an incision in the paper or cardboard, and insert the microscope slide (or the light guide). Note, that light is guided in the x- and y-directions. If the slab is wide in x-direction and narrow in y-direction, as shown in Figure 8.15, then light guiding will be most apparent in y-direction, whereas in x-direction light will be expanding until it hits the edge of the glass and will either be reflected or out-coupled. The principle that governs light guiding is total internal reflection.
2. Out-coupling of guided light by scattering—As apparent in part 1 of this experiment, guided light will be coupled from the end of the slab waveguide. Light can also be coupled from the surface of the waveguide by introducing a "defect" to the surface of the waveguide. Add a defect on one

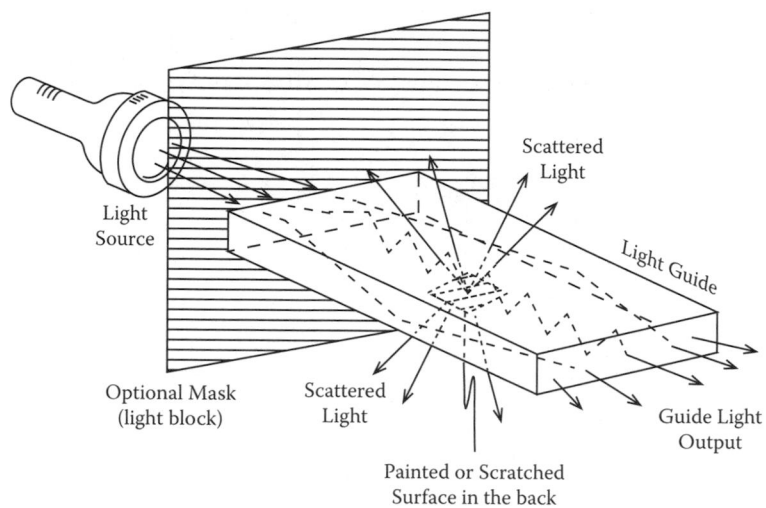

**FIGURE 8.16**  Light out-coupled due to addition of a defect, such a paint, a diffuse tape, or scratching the surface.

of the faces of the slab waveguide by (see Figure 8.16) using one of the following methods:

- White paint,
- Diffuse tape
- Scratch a small area on the surface of the glass with sandpaper

Now repeat the previous test discussed in part 1, as shown in Figure 8.16 and observe the scattered light from both sides of the light guide. The defect causes light to scatter out in both directions. Note that if white paint is used as the defect and if the paint layer is too thick, then scattered light will be visible only from the opposite side.

Both of these experiments should be performed in a dim-lit room for best results.

Photograph of light guiding using a slab waveguide is shown in Figure 8.17a, where a microscope slide is used as a slab waveguide and an LED flashlight is used as the light source, coupled to the right end of the microscope slide. Light output from the left end to of the slide is the guided light. When paint is introduced to the back side of the slide (shown as three dots in Figure 8.17b), light is scattered and can bee seen from the front side.

Variation of these experiments is to observe light guiding in a slab while bending, as shown in Figure 8.18 and Figure 8.19. Observe no light guiding is maintained even while the slab is bent, as shown in Figure 8.19.

### EXPERIMENT 8.2: LIQUID LIGHT GUIDE

Purpose: To familiarize with light guiding concepts
Related chapter/section: 8.3 Planar Waveguides and Integrated Optics
Related simulations: N/A

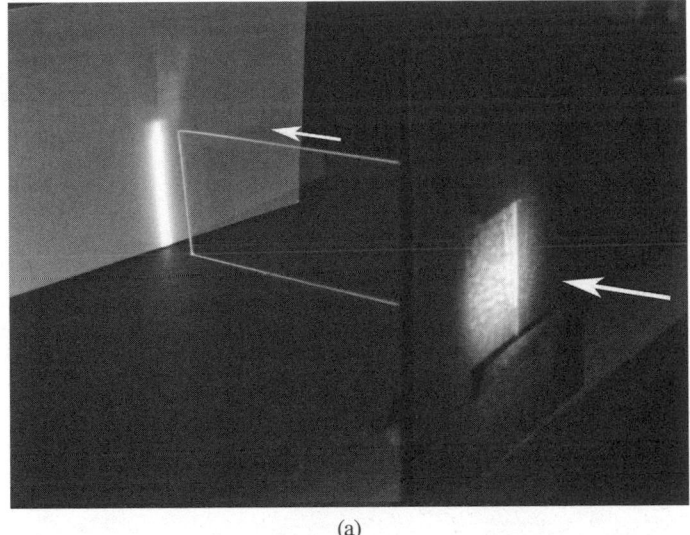

(a)

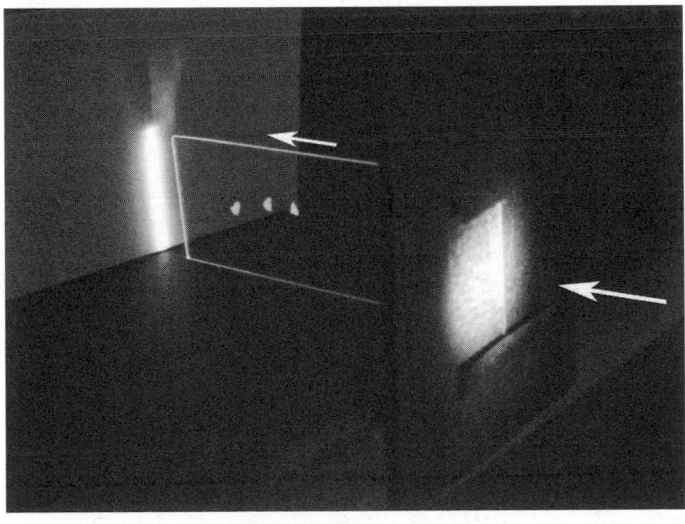

(b)

**FIGURE 8.17**   (a) Illustration of light guiding in piece of slab glass 1 mm thick, as illustrated in Figure 8.15. (b) Light guide with three dots of paint one side that causes light to scatter, as illustrated in Figure 8.16. Arrows indicate direction of light propagation.

Materials needed:

- Clear plastic bottle with flat sidewalls
- Flashlight or laser pointer
- Light block (optional): Foil or black cardboard
- Water

Caution: Observe laser safety rules when using a laser diode.

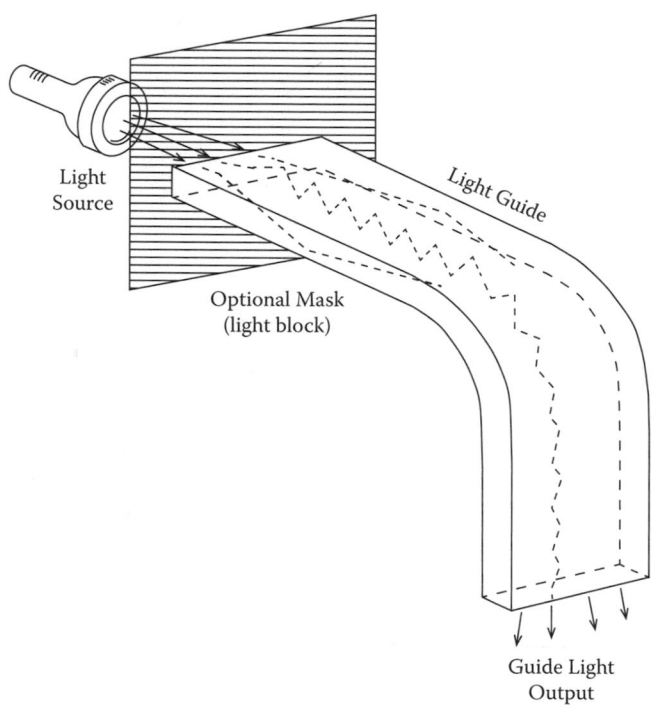

Light Source

Light Guide

Optional Mask
(light block)

Guide Light
Output

**FIGURE 8.18**   Light guided in a bent slab.

### EXPERIMENT

The basic apparatus is shown in Figure 8.20. Create a hole on one side of the plastic bottle for water to flow. Fill the bottle with water while tilting to ensure that water does not flow out of the hole. Place the water bottle on the side of a sink where water can flow while blocking the water outlet. From the opposite side of the hole direct a laser pointer such that laser spots coincides on the hole where water is flowing. The laser propagation angle should be nearly normal to the side-wall of the bottle, as shown Figure 8.20.

While the laser pointer is still pointed as described, let the water flow, and note how light is guided in the flowing water. This can be observed either by placing a white plate in the water path or by placing the palm of the hand in the water path. Mover the plate in the water path and notice how laser light is guided by the water.

The same experiment can be repeated using a flashlight as a light source. However it might be easier to see the effect by blocking the light input area of the water bottle with a cardboard or aluminum foil, but leave an opening opposite to the water hole so that light can go through.

Light guiding phenomenon is best observed if the experiment is performed in a dim-lit room.

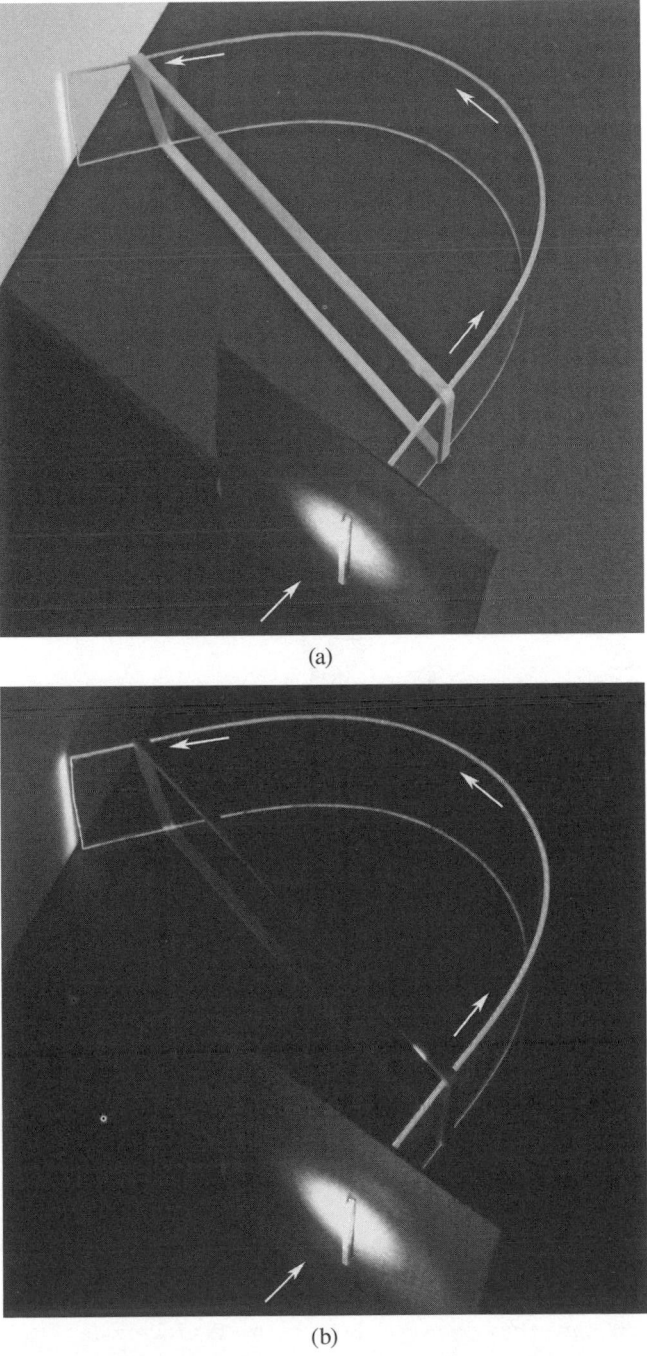

(a)

(b)

**FIGURE 8.19** Demonstration of light guiding in a bent slab waveguide made up of a piece of acrylic approximately 1 mm thick and 2.5 cm wide. Arrows indicate light path. (a) Setup with room lights on. (b) Illustration of light guiding with room lights off. Note that no light is completely trapped in the waveguide, however some light escapes the unpolished edges of the waveguide.

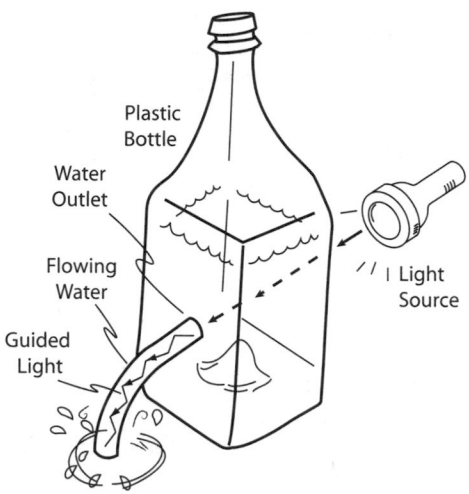

**FIGURE 8.20**    Apparatus to demonstrate light guiding by a liquid. A plastic bottle with flat walls and a hole on one side is filled with water. A laser pointer or a flashlight illuminates the hole, and light guiding is observed in the flowing water column. (Figure drawing by Harout Yacoubian.)

## REFERENCES

1. Jenkins, F. A., and H. E. White. 1976. *Fundamentals of Optics.* 4th ed. New York: McGraw-Hill Book Company.
2. Jones, W. B., Jr. 1988. *Introduction to Optical Fiber Communication Systems.* New York: Holt, Rinehart and Winston.
3. Gagliardi, R. M., and S. Karp. 1995. *Optical Communications.* 2nd ed. New York: John Wiley & Sons.
4. Miller, S. E., and I. P. Kaminow. 1988. *Optical Fiber Telecommunications II.* Boston: Academic Press.
5. Tamir, T., E. Garmire, J. M. Hammer, et al. 1975. *Integrated Optics.* New York: Springer-Verlag.
6. Yeh, P. 2005. *Optical Waves in Layered Media.* 2nd ed. New York: Wiley Interscience.
7. Okoshi, T. 1982. *Optical Fibers.* New York: Harcourt Brace Jovanovich.
8. Marcuse, Dietrich. 1991. *Theory of Dielectric Optical Waveguides.* 2nd ed. New York: Academic Press.
9. Yariv, A., and P. Yeh. 2002. *Optical Waves in Crystals: Propagation and Control of Laser Radiation.* New York: Wiley Interscience.
10. Kapany, N. S., and J. J. Burke. 1972. *Optical Waveguides.* New York: Academic Press.
11. Nishihara, H., M. Haruna, and T. Suhara. 1989. *Optical Integrated Circuits.* New York: McGraw-Hill Book Company.
12. Lee, D. L. 1986. *Electromagnetic Principles of Integrated Optics.* New York: John Wiley & Sons.
13. Binh, L. N. 2011. *Guided Wave Photonics: Fundamentals and Applications with MATLAB® (Optics and Photonics).* Boca Raton, FL: CRC Press.

# 9 Optics, Electronics, Software, and Applications

## 9.1 COMBINING OPTICS, ELECTRONICS, AND SOFTWARE

Many modern optical systems include a combination of optics, electronics, software, and/or firmware. Because each of these portions of the system is interrelated with the other, understanding the capabilities of all three is crucial for an optimal design. Additionally, many portions of the current optical systems utilize digital electronics and microprocessors to replace some of the functionality that used to be performed optically or using analog electronics.

For example earlier spectrometers that analyzed light transmission utilized a reflective chopper wheel to obtain a reference signal from a blank (e.g., a glass) and another through the sample under test (e.g., a glass coated with the material under test). The signal from two photodetectors was subtracted using analog electronics. With the current spectrometers that utilize solid-state detector arrays and digital electronics, this procedure has been simplified, and similar results are achieved with much smaller apparatus and at much rapid rate. First, a scan of through a reference sample is obtained (e.g., a glass substrate), saved in the spectrometer as background, and then the sample is scanned, and transmission through the sample is displayed in real time, referenced to the glass substrate that the sample material is coated on.

Another example is optical image processing, where earlier work on correlation and filtering was performed optically because of the large set of image data and the need for computationally intensive processing. Today much of the processing is performed using digital processing electronics, while optics is used primarily for image collection.

The advantage of using digital electronics and software is the ability to upgrade and optimize without the need for major hardware changes. This is why it is ever more important to have a good understanding of the capabilities of all three: optics, electronics, and software/firmware/processing.

For example, if designing a system that is used to perform three-color measurement on a print based on red, green, and blue (RGB) light-emitting diodes (LEDs), one option is to utilize an illumination system that is designed to maintain constant illumination of all three bands (RGB). This could be prohibitively expensive. If possible, a referencing system may be employed as shown in Figure 9.1. A known stationary reference can be placed in the system such that the response can be simultaneously measured, and the signal can be calibrated. For example if the red and blue LED intensities shift with respect to the green LED, then the print being tested will appear more green than it actually is. By dividing the raw data with the

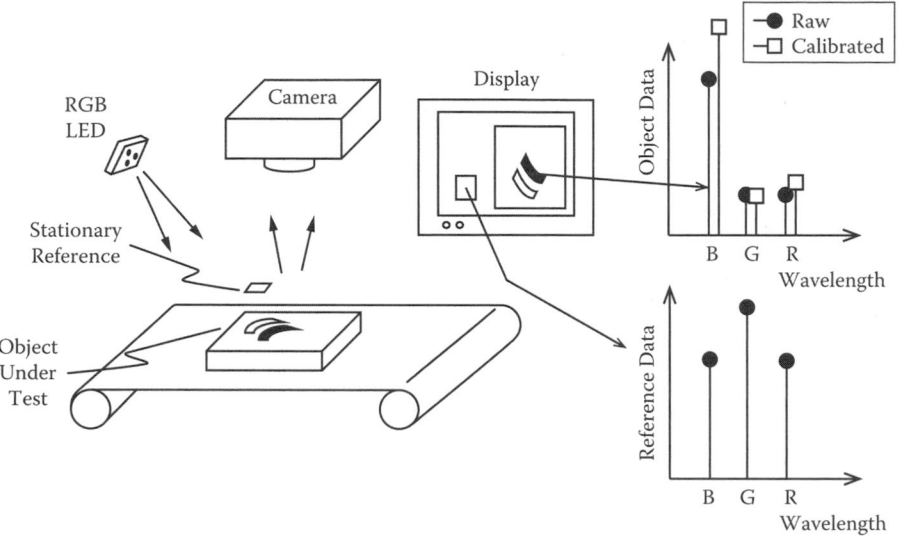

**FIGURE 9.1** Example of a system that employs a reference target to compensate the amplitude variation of an RGB light-emitting diode. Without using a referencing scheme, emission controlled and long-term stability LEDs have to be used to avoid errors in measurements. Using a reference target, and by real-time calibration, lower cost LEDs can be used.

reference signal, the LED color shift can be compensated. Therefore whenever possible, digital electronics can provide a low-cost alternative than using very expensive precision components.

Another example of where electronics can be used to compensate for optics involves resolution of the optical system. When designing an optical system, it is best to match the optics with a required resolution of the system. However, if an exact match is not available, then one must make a choice on what components to use. If one set of optical components (e.g., lenses, camera) results in a resolution that is lower than the desired specifications, then this should be avoided because higher resolution data/image cannot be recreated from a single set of low-resolution data. If, however, multiple data sets (e.g., multiple images) can be taken from a moving target, postprocessing can result in a higher resolution image. On the other hand, if component choices produce an image that is too sharp and a file too large, than the image can be down-sampled (blurred) and file size reduced. Again this is a task that needs to be addressed keeping in mind the available electronics and processing power to perform the postprocessing. Furthermore, the algorithm to use for down-sampling has to be within the processor's capability. Therefore optics, electronics, and software should work together to come up with an optimum solution.

In summary, to start the design of an optical system, it is best to start with specifications in mind, and survey available choices for optics, electronics, processors or computers, firmware, and software. Performing preliminary design work, or even some simulations, will save a lot of time and effort in the future. As described in the preceding examples, if a function can be achieved using digital electronics and software using some referencing/calibration scheme, it may result in a more robust, upgradable, and low-cost system than using precise but fixed components.

## 9.2 SEPARATING OPTICAL AND ELECTRONICS EFFECTS

One of the key aspects of analyzing optical systems is to separate the optical effects from other effects, such as effects due to electronics or software. In many of the modern optical systems, the effectiveness and correct operation depends on optics, electronics, and software and algorithm of the system. Therefore, understanding each of the sections of the system is crucial for designing, optimizing, and trouble-shooting. To optimize or troubleshoot the optical portion of a system, it is important to separate the optical effects from other effects.

An example is to troubleshoot high background noise in an imaging system. To check if this is caused by optics or electronics, the first step is to turn off the light source and measure the noise floor. Compare this to the noise floor with the light source on. If the noise floor is high when the light source is on, then it could be either optical or due to current leakage from the source to the detector. Optical noise could be caused by reflection from a surface or scattering from inside walls of the sensor housing, to name a few. Therefore, separating optical and electronic phenomena through a series of tests will help better understand the source of the problem.

Another example is that an image from a scanner or a camera system does not look sharp. Again this could be optical, electronic, or software or firmware related. A good starting point is to use a resolution test target, such as periodic lines, as described in Chapter 7. If timing of the image capture system can be controlled by the clock rate of the detector array drive electronics, then slowing the clock and recapturing the image reveals if the problem is due to optics or electronics. If the image looks sharp at slower clock rates, then it could mean that the detector response time or the detection electronics are the potential limiting factors. If, however, the images look blurred at any speed, then it could mean an optical problem, such as a misalignment of one of the optical components or a component with a limited numerical aperture.

Another example is testing high-speed optoelectronic systems. For example when working with sensing or communication systems that involve radio frequency (RF) signals (such as signals in the GHz range), it is not uncommon to observe a signal that is due to leakage of the RF source to the detection system. The first test is to disconnect the light source to see if the signal is still observed. If so, various grounding techniques can be employed, including RF shield with metal enclosures to shield the detection electronics from the external electric field. Again a series of optical and electronic tests can help isolate the problem.

## 9.3 APPLICATIONS

This section is a brief summary of examples of applications and relevant topics, given in Table 9.1 and Table 9.2. The readers may find an application that is similar to the work they are doing or plan to do, and be directed to the most relevant chapetrs.

**TABLE 9.1**

**Example Applications and Relevant Chapters**

| Example Applications, Tasks, or Challenges | Relevant Chapters |
| --- | --- |
| • Choosing a light source for variety of applications | • 2: Light Sources |
| • Choosing a detector for visible or near infrared imaging | • 3: Light Detection |
| • Designing a scanner optics | • 4: Manipulation of Light<br>• 6: Geometrical Optics<br>• 7: Imaging Systems |
| • Choosing a microscope | • 4: Manipulation of Light (resolution, numerical aperture, etc.)<br>• 6: Geometrical Optics<br>• 7: Imaging Systems |
| • Choosing a camera for a specific application | • 4: Manipulation of Light (resolution, numerical aperture, etc.)<br>• 6: Geometrical Optics<br>• 7: Imaging Systems |
| • Choosing a telescope for a specific application | • 4: Manipulation of Light (resolution, numerical aperture, etc.)<br>• 6: Geometrical Optics<br>• 7: Imaging Systems |
| • Industrial imaging (e.g., need to purchase the right camera) | • 4: Manipulation of Light (resolution, numerical aperture, etc.)<br>• 6: Geometrical Optics<br>• 7: Imaging Systems |
| • Single point detection for detecting a part on a conveyer belt | • 2: Light Sources (to choose light source)<br>• 3: Light Detection<br>• 4: Manipulation of Light and 6: Geometrical Optics (if lenses and optical components are needed) |
| • Single point detector for color measurement | • 2: Light Sources (to choose light source)<br>• 3: Light Detection<br>• 4: Manipulation of Light and 6: Geometrical Optics (if lenses and optical components are needed) |
| • Distributing light using light guides | • 8: Guided Lightwaves |
| • Fluorescence measurements | • 2: Light Sources (to choose excitation light source)<br>• 3: Light Detection<br>• 4: Manipulation of Light and 6: Geometrical Optics (to understand basics of reflection and refraction, and light propagation) |
| • Designing optical board-to-board interconnect | • 8: Guided Lightwaves<br>• 2: Light Sources<br>• 3: Light Detection |
| • Minimize surface reflections | • 5: Polarization<br>• 4: Manipulation of Light |

**TABLE 9.1 (CONTINUED)**

**Example Applications and Relevant Chapters**

| Example Applications, Tasks, or Challenges | Relevant Chapters |
|---|---|
| • Design multiple lens systems for imaging | • 4: Manipulation of Light<br>• 6: Geometrical Optics<br>• 7: Imaging Systems |
| • Measure stress in the edges of a molded plastic | • 5: Polarization |
| • Design or use fiber optic sensor | • 8: Guiding Lightwaves<br>• 3: Light Detection |
| • Design electronics and choose components for optical storage systems | • 2: Light Sources<br>• 3: Light Detection<br>• 4: Manipulation of Light<br>• 5: Polarization (if polarizing detection is used)<br>• 6: Geometrical Optics |

Table 9.1 describes sample applications and which chapters are the most relevant. These chapters are not specific instructions on how to solve these problems. However, many of the relevant optical concepts are described in these chapters, and familiarizing with them will enhance the ability of the reader to tackle these tasks.

**TABLE 9.2**

**Topics Covered and Relevant Chapters**

| Topic | Brief Description, Terminology, or Where Used | Relevant Chapter |
|---|---|---|
| • Resolution<br>• Modulation transfer function | Measure of the optical system, used in imaging systems | • 7: Imaging Systems |
| • Image distortions | See aberrations | • 6: Geometrical Optics<br>• 7: Imaging Systems |
| • Numerical aperture | (For fiber optics)<br>(For free-space optics) | • 7: Imaging systems |
| • Polarizers<br>• Wave plates<br>• Birefringence | See chapter examples and experiments for various applications | • 5: Polarization |
| • Optical waveguides<br>• Integrated optics<br>• Fiber optics | | • 8: Guiding Lightwaves |

# 10 Optical Sensing

The field of optical sensors is a very broad with many subdisciplines. The purpose of this chapter is to give enough information to get started on the subject and learn the basics, and to demonstrate how optical concepts described in previous chapters apply to optical sensing. Detailed descriptions of each sensing scheme can be found from available literature on various sensing subjects (see, for example, Dakin and Culshaw [1,2]). This chapter incorporates many of the principles used in the previous chapters. Although optical imaging is a form of optical sensing, it is covered in a separate chapter.

Optical sensing systems require a light source, detector, optics, and various optical components, as described in Chapter 1 (see Figure 1.1).

Some sensing systems utilize external light for the sensing scheme. Examples include cameras or systems that detect light emission from a luminescent source. Other sensing systems utilize an internal light source. An example is an interferometric displacement sensor that incorporates a laser as a light source.

Some examples of optical sensors are

- Displacement sensors
- Temperature sensors
- Pressure sensors
- Strain sensors
- Electric-field sensors
- Chemical and biological sensors
- Fluorescence sensors

## 10.1 OPTICAL SENSORS AND SENSING MECHANISM

Many of the optical phenomena described in this book are used for sensing. The following is a partial list of some of the mechanisms used in optical sensors.

### 10.1.1 SENSING CHANGE IN LIGHT INTENSITY

Since all optical sensing utilizes light detection, light intensity measurements are an essential part of all sensing phenomenon. Light intensity measurements can sense changes either due to the sensor configuration (e.g., light transmission changes through a sample) or due to change in the emission of the light spectrum. Depending on the measurement, a proper choice of light source and photodetector (or a detector array) is needed.

## 10.1.2  SENSING CHANGE IN ABSORPTION

Changes in light absorption can be induced by a variety of reasons and can be utilized for sensing by using a light source and photodetector along with optics and electronics. Changes in absorption can be detected by light transmission measurements through a sample. Change in absorption can be caused by a variety of environmental effects, such as chemical exposure or exposure to radiation. Often changes can be wavelength dependent, and therefore either a narrow-band optical filter can be used to isolate a particular wavelength response or a spectrometer can be used to monitor transmission spectra through a sample.

## 10.1.3  CHANGE IN COLOR (WAVELENGTH)

Color change can be measured using a series of detectors with narrow-band optical filters. A color camera is an example of a sensor that detects in three different wavelength bands: red, green and blue. Another method of detecting color change is to use a monochrome camera, one that is responsive to the wavelength band of interest (e.g., using an Si-based camera to detect in the range of 400 to 1100 nm), in conjunction with a series of narrow-band optical filters. For example, a material that is bleached due to environmental conditions can be characterized by measuring the transmission spectrum through that material, and monitoring specific wavelength bands.

Other methods to measure wavelength change are diffraction and spectroscopy. Diffracted light is wavelength dependent, and this can be used for measuring changes in wavelength response. This can be used in free-space geometry as well as in fiber-based sensors. For example, the fiber-grating sensor (discussed in Section 10.2.3) can detect changes in wavelength in the incident wave. Another method is to use a spectrometer. A spectrometer is an instrument that yields light intensity versus wavelength. It is a valuable tool to measure change in wavelength. An example of using a spectrometer is to detect changes in the emission spectrum of a light source, such as a light-emitting diode (LED), due to aging.

## 10.1.4  CHANGE IN REFRACTIVE INDEX

A variety of methods exist that can detect change in refractive index, such as using prism-coupled, fiber optic, or interferometric sensors. Refractive index in passive optical materials can be altered by environmental changes such as stress, pressure, humidity, and temperature. Refractive index of active optical materials can also be altered by magnetic field or electric field. Therefore refractive index measurement can be used to sense external fields and to monitor environmental changes.

## 10.1.5  INTERFEROMETRIC OPTICAL SENSORS

Optical interferometers are used for a variety of sensing applications. Interferometric sensors include the Michelson (see Figure 10.1), Mach-Zehnder (see Figure 10.2),

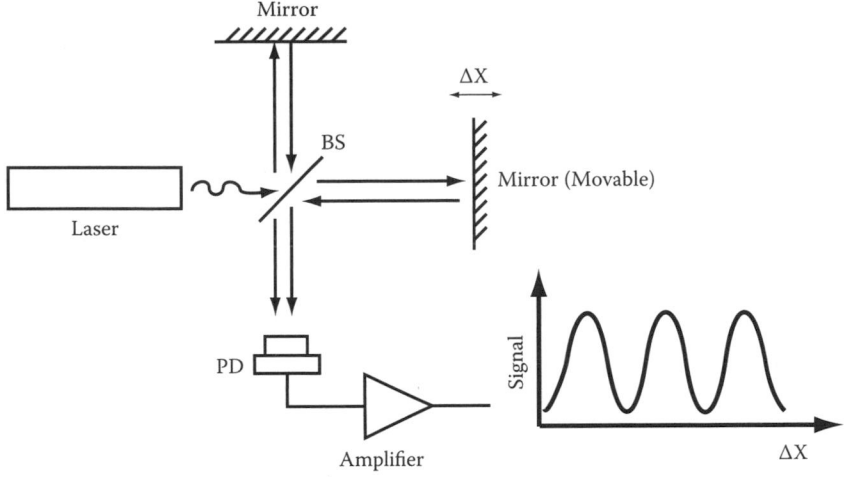

**FIGURE 10.1** Michelson interferometer.

Fabry-Perot (see Figure 10.3), and Sagnac (see Figure 10.4) interferometer sensors. These sensing systems can detect displacement, such as change in optical path length between the two arms of a Michelson interferometer or change in spacing between the two mirrors in a Fabry-Perot interferometer, change in refractive index, or change in wavelength of the light source. A Sagnac interferometer can sense rotation. Interferometric measurements are very sensitive and can measure dimensional changes many orders of magnitude smaller than the optical wavelength. To learn more

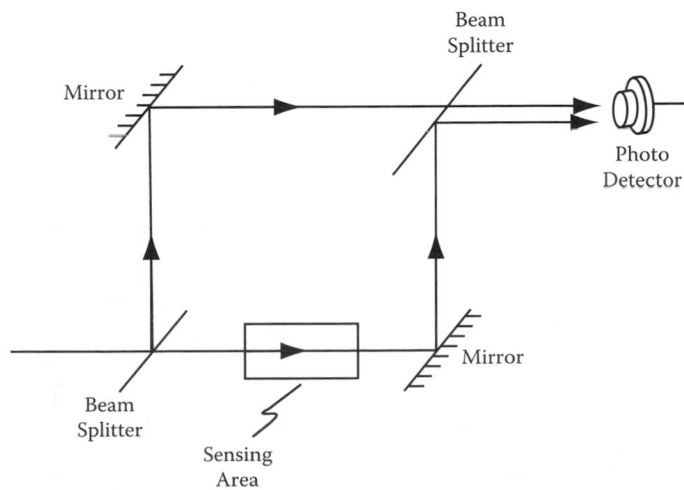

**FIGURE 10.2** Mach-Zehnder interferometer.

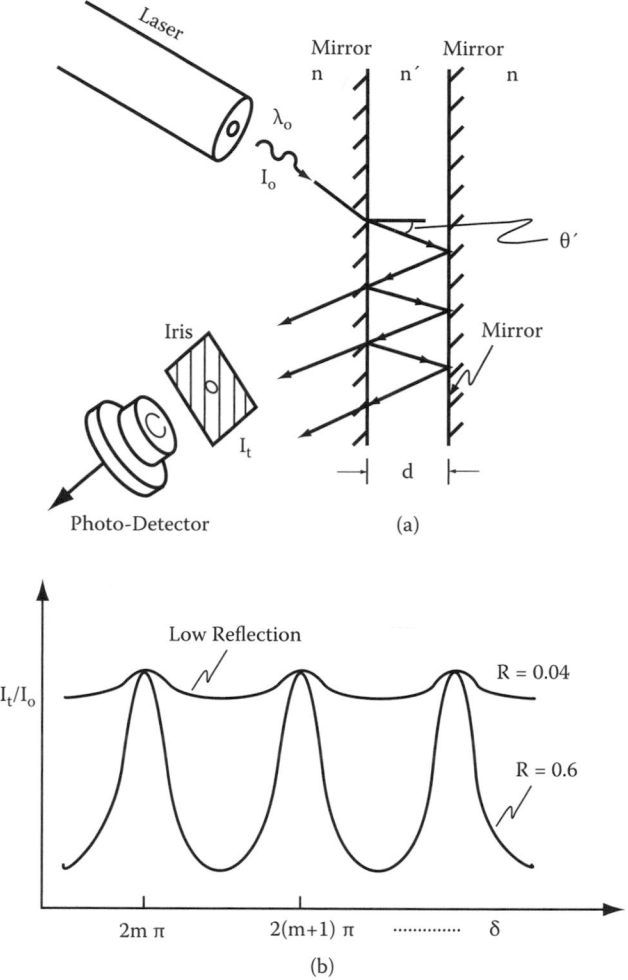

**FIGURE 10.3** (a) A simplified representation of Fabry-Perot interferometer. (b) Interference from Fabry-Perot interferometer with two different mirror reflectivities ($R = 0.04$ and $0.6$).

about interferometry, it is best to start with a basic optics text [3–5], and then migrate to other texts specific to a particular field to which interferometers are applied.

An interferometer measures phase difference between two light waves. The two interfering light waves have to be coherent and generally are from the same source. Therefore interferometers incorporate a method of splitting light and recombining it. One arm of the interferometer is used as reference, and the other is the sensing arm.

In a Michelson interferometer, the interference signal intensity ($I$) at any spatial point can be written as

$$I = \left| \sqrt{I_1} + \sqrt{I_2}\, e^{i\phi} \right|^2$$

(10.1)

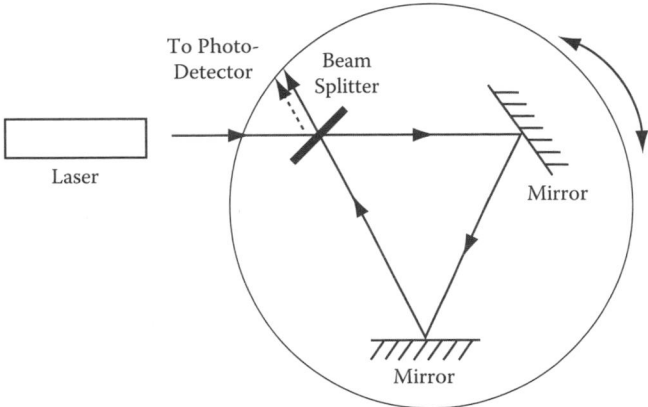

**FIGURE 10.4**   Sagnac interferometer.

where $I_1$ and $I_2$ are the signal intensities from the two arms of the interferometer, and $\phi$ is the phase difference between the two. This equation can also be written as

$$I = I_1 + I_2 + 2\sqrt{I_1 I_2}\, \cos\phi \qquad (10.2)$$

Therefore when $I_1$ and $I_2$ are in phase ($\phi = 0$) the interference signal is high, and when they are out of phase, $I = 0$.

When light in one arm of the interferometer experiences phase shift with respect to the other arm, fringes shift, and the light intensity changes from dark to bright. This phase shift can be the result of various mechanisms, such as moving the reference mirror in a Michelson interferometer, or a refractive index change in one arms of the Michelson or Mach-Zehnder interferometer. This is a very sensitive measurement, because a fraction of a fringe measurement can be detected with a photodetector. Since each fringe movement is on the order of the optical wavelength, measuring fraction of fringe movement results in detection of movement of fraction of optical wavelength or very small changes in the refractive index.

Another example is a Fabry-Perot interferometer shown in Figure 10.3a. The basic structure is comprised of two partially reflective mirrors. Light reflects back and forth from the two mirrors, and some of the light escapes out. The multiple reflected waves interfere, and depending on their phase difference, interfere constructively or destructively, resulting in interferometric fringes. The intensity of the transmitted light, $I_t$, wave is given by

$$\frac{I_t}{I_o} = \frac{1}{1 + F\sin^2(\delta/2)} \qquad (10.3)$$

where $I_o$ is the incident light intensity, and the phase difference, $d$, between transmitted waves is given by

$$\frac{\delta}{2\pi} = \frac{2n'd\cos\theta'}{\lambda_o} \qquad (10.4)$$

where $n'$ is the refractive index in the Fabry Perot cavity, $\theta'$ is the angle of the light ray inside the cavity, and $\lambda_o$ is the light wavelength [5], and $F$ is the coefficient of finesse [3,5], and is given by

$$F = \frac{4R}{(1-R)^2} \tag{10.5}$$

where $R$ is the reflectivity of the mirrors. The higher the reflectivity, the sharper are the fringes as shown in Fig. 10.3b.

Any changes in the parameters $n'$, $d$, $\theta'$ can be detected by observing changes in the detected signal. Fabry-Perot interferometers are used for spectroscopy with high resolving power. One spectroscopy application is used to study spectral lines of a laser source. Fabry-Perot is also the structure of a laser cavity, where the lasing medium is placed in the Fabry-Perot cavity, with one mirror being highly reflective, and the other partially reflective allowing laser light to escape the cavity.

### 10.1.5.1 Some Applications of Interferometry

One application of interferometry is quality inspecting of optical components such a lenses and mirrors. Another application of interferometers is for surface profiling. An example of such a technique for surface measurement is white light interferometry, where fringes from a white light source are obtained using special types of microscope objectives (called interferometer objectives). This method enables precision surface profiling. Interferometry can achieve surface measurements with depth resolution several orders of magnitude smaller than the optical wavelength.

### 10.1.6 Sensing Change in Polarization Angle

Polarization can be used to measure a variety of environmental parameters. One example is measuring stress in a plastic film by measuring birefringence. An example of this is shown in Chapter 5, Figure 5.2, where stress in a plastic film can be measured by placing the plastic film between two cross-polarizers.

### 10.1.6.1 Polarization-Dependent Reflection

An example of polarization-dependent reflection sensing is to detect transparent objects on a surface by detecting difference in polarization-dependent reflection coefficient. This is demonstrated for tape detection in Chapter 5, Figure 5.5. Another commonly used application of polarization reflection is for thin film measurements. Multilayer optical thin films can be measured for thickness and refractive index by using polarization measurements, where reflectance at different angles of incidence and at various polarization orientations are measured, and the data curve fitted to estimate the thin film parameters.

### 10.1.6.2 Polarimetric-Based Electric and Magnetic Sensors

Polarization can be used to measure electric and magnetic fields by use of electro-optic (EO) or magnetooptic (MO) crystals and materials. EO materials respond to electric field, and changes are induced in refractive index at a particular orientation. The change is detected by placing the EO material between cross-polarizers and observing the change in the light intensity incident on the detector. Similarly, for magnetic field sensing, an MO crystal is used in the same configuration.

### 10.1.7 SENSING BY DETECTING CHANGES IN DIFFRACTION ANGLE OR WAVELENGTH

Light diffraction from a periodic structure is dependent on the angle of diffraction and wavelength, namely,

$$2\Lambda sin\theta = m\lambda \tag{10.6}$$

where $\Lambda$ is the grating (or periodic structure) period, $\theta$ is the diffraction angle of the $m$th order, and $\lambda$ is the wavelength of light. Changing the period, $\Lambda$, of the periodic structure changes the diffraction angle. Therefore measuring the diffraction angle can be used to measure dimensional changes to the periodic structure. Similarly, changes to the peak diffraction wavelength can also be used to measure dimensional changes. An example of this is used as fiber sensor to measure strain, as shown in Figure 10.5. A periodic structure is embedded in the fiber via one of the several techniques available (such as etching or laser writing [6]). The grating is optimized for a particular wavelength of light, such that light is diffracted back, and minimum light is transmitted when the fiber is in a nonstressed position. Whenever there is a change in the fiber dimension, diffraction efficiency changes, and less light is diffracted back than in a nonstressed position. When the fiber dimension changes, light intensity measured by the detector increases. The intensity measurement can be converted to strain measurement and can be used for structural health monitoring [2].

### 10.1.8 SPECTRAL SENSING OF TEMPERATURE

As described in Chapter 2, light emission from a blackbody source (such as hollow metallic cavity with a small opening) depends on the temperature of the cavity. Therefore using a detector in conjunction with a narrow band filter can be used as a highly sensitive method of measuring the temperature of a blackbody source. Often multiple sensors with multiple narrow band filters are used for more precise

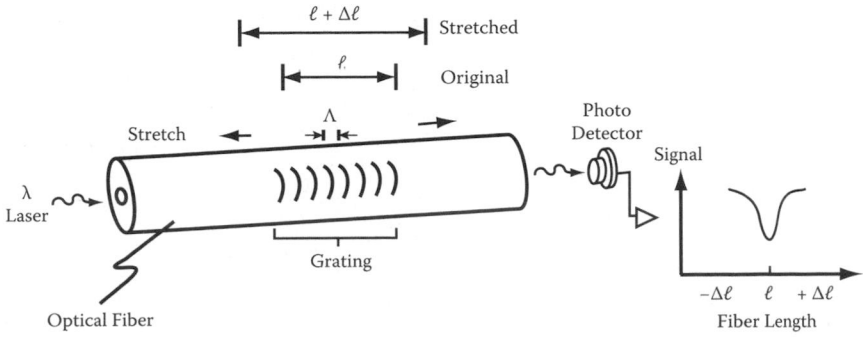

**FIGURE 10.5** Fiber Bragg grating sensing changes to fiber length by monitoring changes in the photodetector signal.

measurements. Current advancements in the spectrometers and reduced cost due to wide use of spectrometers that utilize detector arrays enables even more precise measurement of temperature spectroscopically by observing the blackbody emission curve. These types of techniques are utilized both in a free-space configuration, as well as fiber-coupled configurations. Many of the modern spectrometers offer options of either free-space or fiber-optic coupling of light to the spectrometer.

### 10.1.9 SENSING FLUORESCENCE EMISSION

Some materials when excited by light of a given wavelength emit light at a different wavelength (e.g., excitation with ultraviolet source and emission in blue). When the process is fast, such as in sub-microsecond time scale, it is generally called fluorescence. When the time scale is long (e.g., seconds or minutes), it is referred to as phosphorescence [3]. Fluorescence can be used to characterize materials. Light emission from a fluorescent material can be measured by a number of optical methods. These methods include free-space sensing configuration, fiber optic sensors, or a microscope. When the emitted intensity is very small compared to the excitation source, various techniques are utilized to minimize the excitation light source from saturating the detector. These techniques include designing an angularly selective detection arrangement such that most of the excitation light is blocked, utilizing narrow-band optical filters that only transmit the fluorescent wavelength and blocks the excitation light (see, for example, Figure 10.6), or using a modulation scheme, where the excitation light is modulated, and a lock-in amplification scheme is used as described in Chapter 3. Often a combination of few of these techniques is utilized to measure fluorescence.

### 10.1.10 SENSING FLUORESCENCE LIFETIME

Fluorescence lifetime can be used to characterize fluorescent materials. When a fluorescent material is excited (e.g., a light pulse) light emission will decay over time.

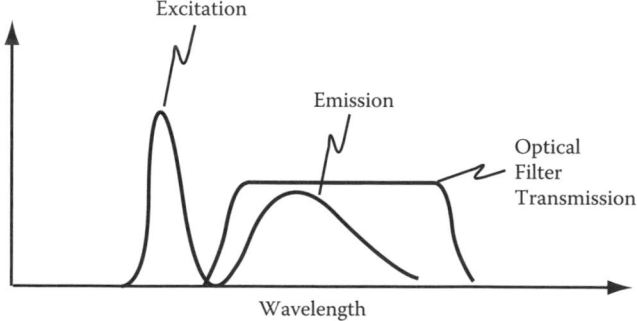

**FIGURE 10.6** Fluorescence excitation and emission spectra, and the use of optical filters to block the excitation light from saturating the detector. Excitation and emission amplitudes and spectral width are not to scale. Often the emission is much smaller than excitation.

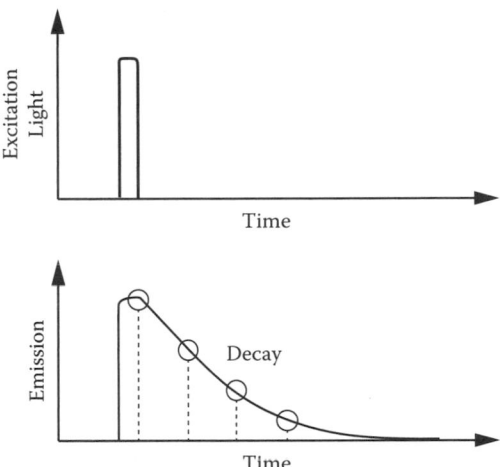

**FIGURE 10.7** One method of fluorescence lifetime measurement used in fluorescent lifetime imaging based on time-based measurements. After the excitation pulse is turned off, emission is detected at different time intervals and fluorescence lifetime is obtained.

The rate of decay is often used to characterize materials. Light intensity, $I$, decays over time, $t$, starting at the initial light emission amplitude, $I_o$, at the rate $\tau$:

$$I(t) = I_o e^{(-t/\tau)} \tag{10.7}$$

Often there is more than one phenomena occurring at the same time, and there will be multiple decay rates, $\tau$. Measuring this decay is used extensively in biology.

In some applications, single point detection is sufficient. In other applications, particularly in biology, microscopic imaging is desired. This is called fluorescence lifetime imaging microscopy [7], where the fluorescent material is placed under a microscope, and various excitation and detection apparatus are added to the microscope setup.

To measure time decay, various techniques can be used. One method is excitation with a pulsed illumination, and detection after light is turned off at various time intervals, as shown in Figure 10.7.

Another method is by detecting phase modulation, namely, by modulating the excitation source, and detecting the phase delay of the emission.

## 10.1.11  HOLOGRAPHY-BASED SENSORS

Holographic sensors achieve very sensitive measurements. Many of the holographic methods employ photographic film, which requires postprocessing. Other holographic techniques utilize a camera instead of a photographic film. They are less sensitive but can perform real-time measurements, such as speckle holography [8].

One of the holographic techniques adapted for sensing was double exposure holography. The basic principle of double exposure holography is to create a hologram, namely, an amplitude and phase recording in a film with an object at a given point in time, and then repeat the exposure at another time. If there is any change in the object dimensions, and when the recorded hologram is illuminated, the resulting image will contain dark and bright lines (coarse fringes) overlapping the image. One application example is to inspect airplane tires. The first exposure is taken with the tire at one pressure, and the second exposure is taken by slightly increasing the tire pressure. Normal areas of the tire will show uniform and coarse fringes. If there is a weak spot that changes more rapidly than the rest of the tire, then fine fringes (typically circular fringes) are observed around the weak spot. Various holographic measurement techniques are described in the literature and in books on holography [8].

### 10.1.12 SURFACE PLASMON BASED SENSORS

One particular type of sensing configuration is called a surface plasmon sensor. It is used to measure small changes in refractive index induced in a sample [2] due to changes in surface chemistry. Surface plasmon (SP) resonance is achieve by using a thin metal film, such as 50 nm thick silver layer, in conjunction with a high-refractive index prism, as shown in Figure 10.8. Surface plasmon resonance occurs at a dielectric–metal interface when the momentum of the photon in the film plane matches the surface plasmon momentum, $k_{sp}$. Namely,

$$k_{sp} = \frac{\omega}{c} \sqrt{\frac{1}{\varepsilon_s} + \frac{1}{\varepsilon_m}} \tag{10.8}$$

where $\varepsilon_s$ and $\varepsilon_m$ are the dielectric constants of the sample and metal layers, respectively. The resonance occurs at an angle $\theta_{sp}$ where reflected light is minimized, as shown in Figure 10.8. At this angle, the electric field amplitude of the evanescent field (namely, the nonpropagating portion of light that extends out of the prism) is enhanced by two orders of magnitude, compared to if no metal film was used and the

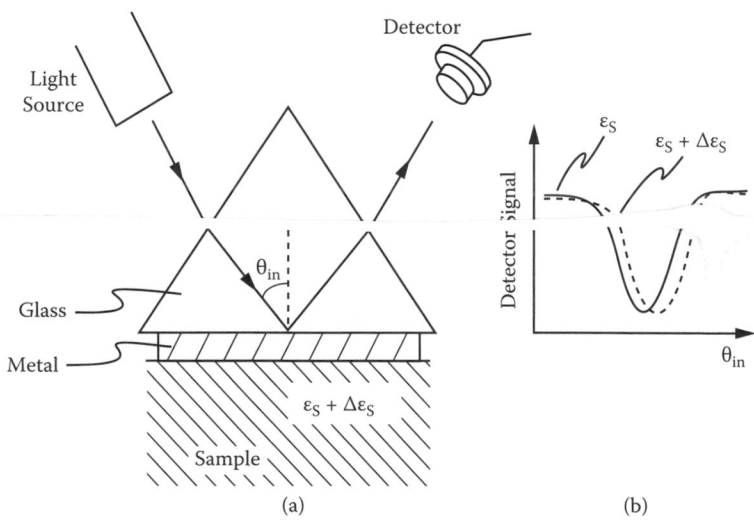

**FIGURE 10.8** Surface plasmon sensor.

sample film was coated directly onto the prism. Any changes to the sample causes shift in the resonance angle, as shown in Figure 10.8, which is detected by the photodetector. This setup is also used in an imaging configuration, where instead of a single element photodetector, a camera is used. The incident light is collimated, and the prism is positioned at the resonance angle where reflection is minimized. When there is a change in the dielectric constant, $\varepsilon_s$, due to changes in the properties of the sample at a particular spatial position, then this area appears brighter. Therefore such a setup can be used to localize surface reactions. SP has been used for chemical sensing, sensing presence of gases, and studying immunoassay reactions, to name a few [1,2].

## 10.2 FIBER OPTIC SENSORS

Since the advent of fiber optics, a large number of sensors based on fiber optics and guiding lightwave principles have been introduced. Fiber optic sensors have become popular due to the advances in fiber optic and electro-optic components and advances in the fiber optic communication industry. Components used for communication applications were also used in sensor industries. Mass production resulted in reduced prices and quality components, which enabled fiber optic sensors to displace nonoptical sensors used to measure displacement, rotation, stress and strain, current, pressure, temperature, electric field, magnetic field, and chemical and biological sensing to name a few. A basic fiber optic sensor system generally consists of a fiber optic sensor head, and lasers, detectors, and electronics. Detailed descriptions of different types of optical sensors can be found in a variety of texts (see, for example, a book by Culshaw and Dakin [1,2]).

### 10.2.1 INTENSITY DETECTION FIBER OPTIC SENSORS

An example of an intensity-based fiber optic sensor consists of two fibers with polished end faces facing each other at close proximity. Light exiting one fiber is coupled to the other fiber. One fiber is mounted on a stationary position and another in a movable position. Light coupling between the two fibers is dependent on the acceptance angle of the fiber (the numerical aperture of the fiber) and the distance between two fibers. When distance between the two fibers change due to vibration or strain, the amount of light coupled between two fibers changes, and the signal detected by the photodetector placed at the end of the receiving fiber changes.

Another example is an optical fiber with the end polished at an angle such that light is reflected via total internal reflection (TIR, as described in Chapter 8) and another section of the fiber is polished and mirrored, as shown in Figure 10.9. If the surrounding medium is low refractive index, such as air, then light reflects from the polished endface via TIR, and is reflected back from the mirrored section, and reflected again via TIR and travels back and is detected by a photodetector, which measures a high signal. If the surrounding medium changes, such as if the fiber is dipped in a liquid, TIR conditions change, and light can be coupled out of the fiber. Less light is received by the photodetector (PD) and the photodetector signal drops. This method can be used as a liquid level sensor as well as a sensor to measure changes in the refractive index of a liquid.

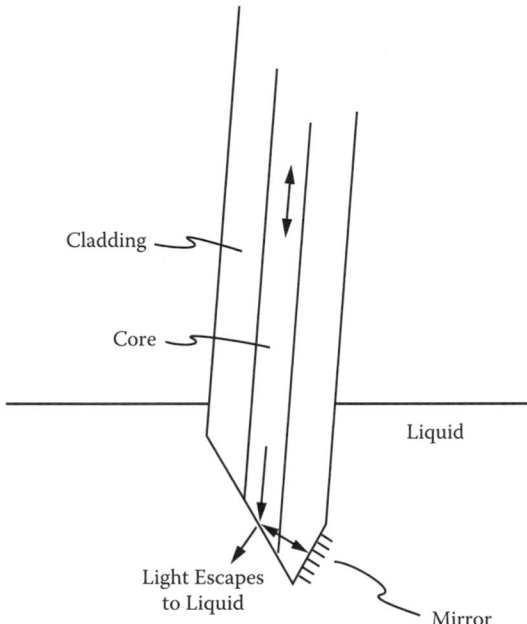

Cladding

Core

Liquid

Light Escapes
to Liquid

Mirror

**FIGURE 10.9** Optical fiber sensor with the end polished at an angle such that light is reflected via total internal reflection (TIR) and another section of the fiber is polished and mirrored. If the surrounding medium is low refractive index, such as air, then light returns due to TIR. If the fiber is dipped in a liquid, light can be coupled out of the fiber in the polished side, and the return signal is reduced.

Another example of a TIR sensor is shown in Figure 10.10, where light from a fiber is coupled to a prism, and coupled back to another fiber via total internal reflection [2]. When the prism is dipped in a liquid, light is coupled out to the liquid and the photodetector signal is reduced.

## 10.2.2  EVANESCENCE FIELD FIBER OPTIC SENSORS

Several types of evanescent field fiber sensors exist. Evanescent field is the optical field that is outside the core of the fiber and extends into the cladding. This optical field is nonpropagating, and it decays rapidly away from the core (the center high-refractive index section of the fiber). However, when the evanescent field is disturbed, light can couple out of the fiber core. This is the reason why optical fibers have a cladding, namely, to avoid light being coupled out of the core due to changes in the environment. For optical sensing applications, however, the cladding (the outer low-refraction index section of the fiber) size can be reduced, or the cladding can be removed, or a large bend can be introduced to the fiber such that the evanescent field can "see" the surrounding environment. These types of sensors can be used to measure changes in temperature, pressure, strain, and electric field, to name a few.

One of example of an evanescent field fiber optic sensor is a fiber pair where their cores are placed at close proximity (e.g., if the cladding is polished). When the two

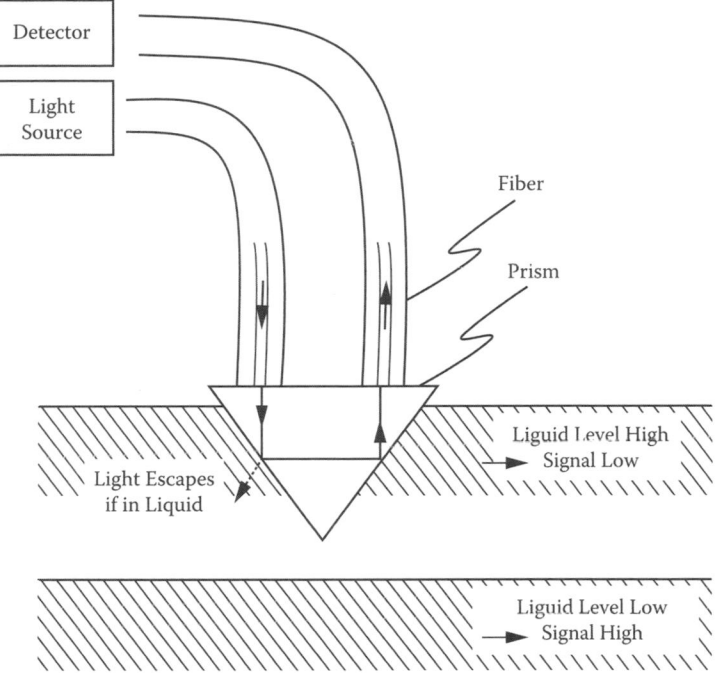

**FIGURE 10.10** A sensor based on fiber/prism total internal reflection. When the prism is dipped in a liquid, light escapes and the photodetector is reduced.

fibers are placed at close proximity, light from one fiber can be coupled to the second fiber. The amount of light coupling is dependent on the distance between the fibers, and it can be used as a sensor that can measure very small changes in distance.

Another type of an evanescent field fiber optic sensor is the microbend optical sensor, where a small radius is introduced such that the conditions for total internal reflection are not met (see Chapter 8), and light couples out of the fiber. Light transmission through the fiber can be modulated by microbending the fiber. Microbend-based fibers have been used in applications such as sensing pressure and vibration.

Another application is an electric-field sensor achieved by a fiber with the cladding removed and the core exposed. Adjacent to the core an electro-optic slab waveguide is placed. This layer can consist of an EO crystal [9] or an EO polymer, as shown in Figure 10.11. Because of close proximity of the EO waveguide and the slab waveguide, part of the light is coupled into the slab waveguide and eventually escapes the waveguide. The amount of coupling is dependent on the refractive index of the slab waveguide. The refractive index of an EO material is altered by applying electric field across the EO slab waveguide. Therefore this sensor can detect external electric field.

### 10.2.3 FIBER-GRATING SENSORS

A diffraction grating (e.g., periodic lines) can be created by optically writing the grating directly into the core of the fiber. An example is a diffraction grating written

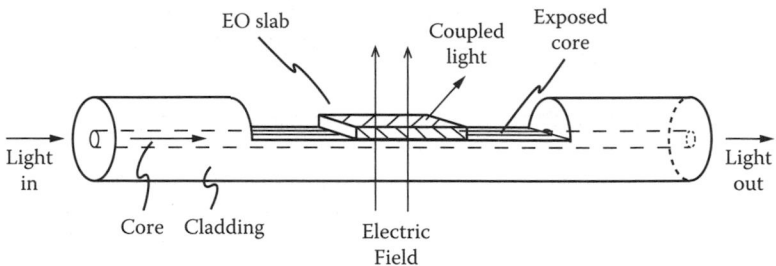

**FIGURE 10.11**  Electric field sensor comprised of an optical fiber with cladding removed and a slab waveguide comprised of an electro-optic material placed adjacent to the fiber core. The amount of light coupling out of the fiber to the slab waveguide depends on the amplitude and orientation of the external electric field.

in a germanium-doped optical fiber using a high power laser. One of the applications of a fiber-grating sensor is for structural health monitoring. The intensity of the diffracted light depends on the source wavelength and the period of the grating. When such a fiber is embedded in a structure, and changes in the structure cause the fiber to expand, then the diffraction conditions change. If the wavelength is fixed and the period changes, then intensity of light diffracted back from the grating changes. This change can be monitored to determine any changes to the structure (see Figure 10.5).

### 10.2.4  MICHELSON AND MACH-ZEHNDER INTERFEROMETRIC FIBER OPTIC SENSORS

Interferometric sensors have much higher sensitivity and dynamic range than non-interferometric sensors. Michelson (Figure 10.12) and Mach-Zehnder (Figure 10.13) fiber-based interferometers measure the difference in phase between two arms of the

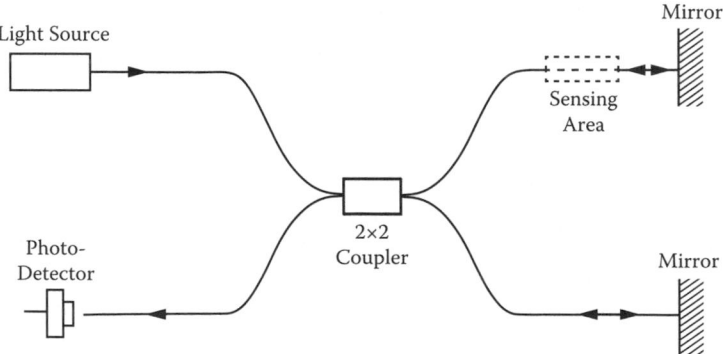

**FIGURE 10.12**  Fiber-optic-based Michelson interferometer sensor.

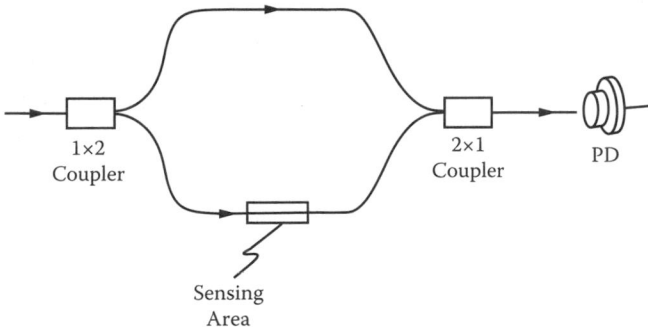

**FIGURE 10.13**  Fiber-optic-based Mach-Zehnder interferometer sensor.

interferometer. Any changes that induce phase change, such as strain, displacement, and refractive index, cause the signal at the detector to change. For example, if one arm of the interferometer contains a fiber embedded in a structure, and when there is a change in the structure that causes change in the fiber length (e.g., a bend or stretching), then the signal at the detector changes. These types of sensors are often used in many applications such as for structural health monitoring. Figure 10.12 and Figure 10.13 show a simplified version of the fiber-sensing scheme. However there are many more components that go into this system, including modulators and demodulators [1,2].

### 10.2.5  SAGNAC INTERFEROMETER FIBER OPTIC SENSORS FOR ROTATION SENSORS

Sagnac interferometer fiber optic sensors (Figure 10.14) are primarily used for rotation sensing. Many of the traditional gyroscopes are being replaced by fiber-optic gyroscopes, which do not require mechanical moving parts. Phase shift is induced by a rotating fiber loop, which is proportional to the rotation rate of the fiber loop as shown in Figure 10.14. Other variations of fiber-optic gyroscopes utilize similar Sagnac interferometer scheme with modulation demodulation components [2].

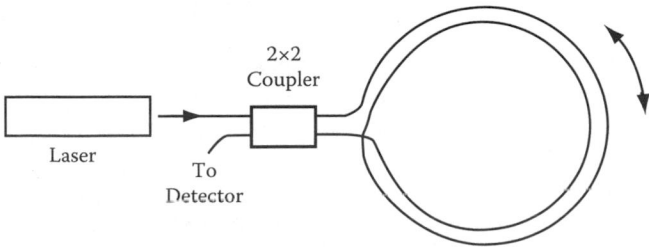

**FIGURE 10.14**  Fiber-optic-based Sagnac interferometer.

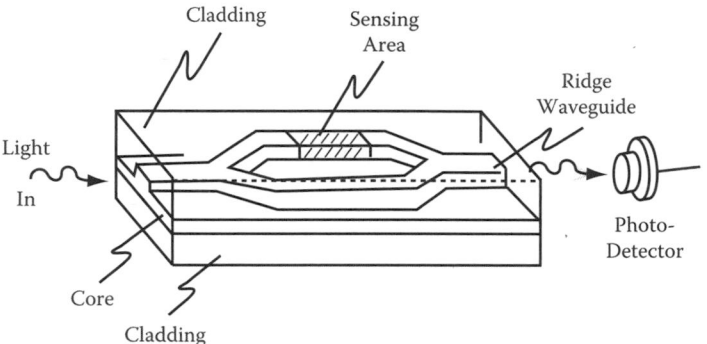

**FIGURE 10.15** Integrated optic electric-field sensor that utilizes active electro-optic material in one arm of the interferometer. An external electric field induces small change in the refractive index in the area with the EO material and interference signal changes.

### 10.2.6 INTEGRATED OPTICAL SENSORS

Integrated optical sensors utilize planar structures for controlling direction of light propagation. They are fabricated using a variety of techniques. One common method is utilizing tools that are used for semiconductor wafer manufacturing. These tools and methods are used to pattern waveguides using optical lithography, to grow and etch materials to shape the waveguides. Integrated optical devices use the principle of light guiding as described in Chapter 8. Both passive and active integrated optical devices have been used for sensing. Active integrated optical sensors have active material as part of the device. One example is an electric field sensor that utilizes active electro-optic (EO) material in one of the arms of a Mach-Zehnder interferometer as shown in Figure 10.15, where an external electric field induces small change in the refractive index in the area with the EO material, interference signal changes. This change is detected by the photodetector.

## 10.3   IMAGING SENSORS

Many of the sensors described in this chapter can be expanded to imaging or to generate two-dimensional rather than a single point data. Advancements in camera systems, high-speed electronics, and computing made it possible for many of the single point sensors to be converted to imaging systems.

One example of imaging sensors is the imaging spectrometer, where the spectrum of an image is obtained using an imaging apparatus that is coupled to a spectral measurement apparatus. These techniques are called hyperspectral imaging, where a three-dimensional data is obtained, two-dimensional for image in the x-y space, and a third dimension would be wavelength. A coarse variation of this is multispectral imaging, where the wavelength bands are not as narrow as a hyperspectral imaging, and the images are taken either by multiple camera systems or using a single camera in combination with optical filters.

Another application of imaging is microscopic sensing. Many of the current biological testing systems incorporate microscopic imaging apparatus, such as fluorescent microscopy.

## 10.4 STARTING POINT TO DESIGN OR CHOOSE A SENSING SYSTEM

The first step is to determine and outline the goals, and specify the requirements and desired specifications. The next step is to see if there are sensors or systems out there that can be purchased and if it is within the correct specifications. If such sensor is not available, an alternative is to create the sensor from the ground up. The specifications include what need to be detected and at what precision and range. Also the available size, weight, and ruggedness are some of the parameters to consider. In reality however, neither the sensor specs nor the methods may be very apparent at first. Therefore the best approach may be to tackle the problem from both ends. Preliminary specs may be developed as a general guideline starting with the desired outcome of the project, and some design and experimentation can be carried out to assess some of the potential methods and available components.

As another example, there might be a need for a sensor, but the desired specifications may not be clear because the data processing or algorithm is not developed nor is it known how to perform. In this case, again tackling the problem from both ends would be beneficial. For example, the end user (the algorithm person) will perform preliminary calculations or simulations to estimate desired or estimated specs, and the sensor designer (if different from the algorithm developer) will utilize these specs to develop the sensor. Next, the algorithm developer and the sensor developer will agree on the specifications and continue with the design and implementation. This often will eliminate under- or over-designing the sensor system, thus reducing time of development and minimizing potential problem.

## REFERENCES

1. Dakin, J., and B. Culshaw. 1988. *Optical Fiber Sensors: Principles and Components.* Vol. 1. Boston: Artech House.
2. Dakin, J., and B. Culshaw. 1988. *Optical Fiber Sensors: Principles and Components.* Vol. 2. Boston: Artech House.
3. Hecht, E. 2001. *Optics.* 4th ed. Reading, MA: Addison-Wesley.
4. Klein, M. V., and T. E. Furtak. 1986. *Optics.* 2nd ed. New York: John Wiley & Sons.
5. Born, M., E. Wolf, et al. 1999. *Principles of Optics: Electromagnetic Theory of Propagation, Interference and Diffraction of Light.* 7th ed. New York: Pergamon Press.
6. Kashyap, R. 2010. *Fiber Bragg Gratings.* New York: Academic Press.
7. Lakowicz, J. R. 2006. *Principles of Fluorescence Spectroscopy.* New York: Springer.
8. Collier, R. J., C. B. Burckhardt, and L. H. Lin. 1971. *Optical Holography.* San Diego, CA: Harcourt Brace Jovanovich.
9. Tamir, T. (ed.). 1975. *Integrated Optics.* New York: Springer-Verlag.

# 11 | Advanced Experiments

This chapter describes experiments that require more advanced apparatus than what is used in the experiments described at the end of the previous chapters. The reason for not including these experiments in those chapters is to be in line with the book's objective and that is to keep everything to simplicity. However, this chapter takes the reader one step further to perform more complex experiments, using apparatus that are generally available in many technical laboratories.

### EXPERIMENT 11.1: HOMEMADE SPECTROMETER

Purpose: To become familiar with the use of diffraction in spectroscopy
Related simulations: N/A
Materials needed:

- Incandescent or halogen flashlight
- Diffraction grating (on a reflective foil); Alternative: holographic element on a discarded credit card
- Some cardboard and tape and glue to make an enclosure for light blocking
- Camera (e.g. a digital camera)
- Software to convert image data to numbers (e.g., MATLAB®)

#### EXPERIMENT

A grating spectrometer is an instrument that is used to disperse light, namely to separate it to different wavelengths. The data obtained from a spectrometer is a plot of signal versus wavelengths. In this experiment, the spectrometer measures spectral transmission through a sample (e.g., a colored filter). A spectrometer consists of a light source, a component to split the light spectrum (e.g., split white light to "rainbow" colors), a detector or a detector array, and collecting optics. Either a refractive element (such as a prism) or a diffractive element (such as a grating) can be used to split the light spectrum.

A readily available component that can be used as a diffraction grating is a compact disk (CD), because of the periodic pits that are used to record data. One drawback of using a CD is the diffraction pattern is curved (rather than linear), because of the periodic structure (the pits) are on a curved path. Because of this, it is necessary to correct for this. However if a linear grating is used, then the pattern is not curved. An alternative is to use the holographic element on a discarded credit card. Other alternatives are diffractive foil, sometimes found on certain types of chewing gum, wrappers, or decorative paper, or to purchase a low-cost diffractive element available from optics vendors.

The basic setup is shown in Figure 11.1. Light enters through an opening between the light blocks, and is diffracted from the diffraction grating. The camera captures the diffracted light. The reason for using the light blocks in Figure 11.1 is to block light from entering at multiple angles, and limit it to a narrow angle. In an off-the-shelf spectrometer, typically a lens or a concave mirror is used following a

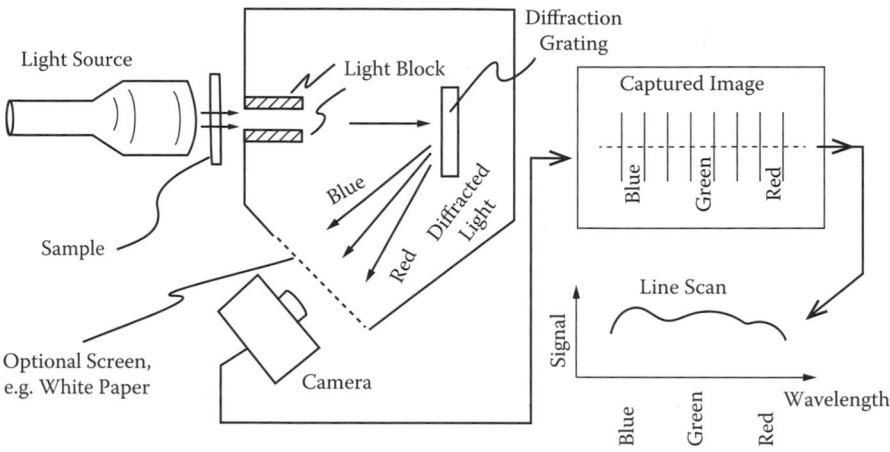

**FIGURE 11.1**   Example of a simple spectrometer using a flashlight, diffraction grating, and a camera.

Single hole or slit, and another lens or concave mirror collect the diffracted light (see, for example, the Czerny-Turner spectrometer [1]). However, for simplicity, here we use light blocks (which can be constructed from two pieces of cardboard), and the camera captures the diffracted light spectrum. If a white paper (or a diffuse screen) is positioned as shown in Figure 11.1, the diffracted light will create a rainbow pattern on this screen, which is then captured by the camera. This experiment should be performed in a dim-lit room or the room light should to be blocked, so that the camera only captures the diffracted light.

The diffraction grating should be rotated until the diffracted light appears on the screen, with the red color farthest from the slit and blue closest. It should be noted that diffraction gratings produce multiple diffraction orders, and the first order is generally the brightest. Therefore the grating should be rotated until the brightest diffraction spectrum is visible on the screen.

To obtain a spectral plot, namely, a plot of signal versus wavelength, the captured image has to be converted to pixel values. Computational software with image reading capability such as MATLAB can read digital image data and convert it into pixel values. A plot of the pixel values for a line in the center of the image of the spectrum (from blue to red) will reveal the spectral response, as shown in Figure 11.1.

The experiment preferably utilizes a monochrome camera. Alternatively, if using a color camera, RGB image pixel values have to be converted to grayscale intensity values in the software.

An alternative to this setup is to use a transmission diffraction grating, but this requires repositioning the screen and camera accordingly.

### DETECTION AND CALIBRATION

Wavelength calibration: Once the location of the components and the camera position are fixed (and the camera optics, such as zoom, do not automatically change), then it is possible to estimate the wavelength from the captured image. To do this, it is best to use several narrow wavelength band sources, such as red, green, and blue light-emitting diodes (LEDs) with known wavelengths, and capture an image while noting the position of diffracted light in the image with each LED. This information can be used to estimate the wavelength of the diffracted light from the captured image. Namely, the x-axis of the plot shown in Figure 11.1 can be converted to wavelength values.

Amplitude calibrated transmission spectrum: To obtain a calibrated transmission spectrum from a sample (such as a colored filter), two sets of data are needed. One is background signal, namely, signal without the sample, and another is with the sample. There are two methods of achieving this: One is to use the setup shown in Figure 11.1 and first capture the image of the spectrum without the sample, and then capture the image of the spectrum with the sample. However, to avoid errors, the apparatus should not move between the two image captures, and the camera autocorrection function should be turned off.

A second alternative is to use half the image to capture the data through the sample, and the other half to capture the reference data. In this case the light blocks shown in Figure 11.1 are constructed such that they form a slit in the out-of-plane direction of the figure. The sample can be positioned such that it covers only part of the slit (i.e., to the top or bottom part). Therefore there will be two spectra on the screen: One representing transmission through the sample and the other is the background.

Regardless of which method is used, two sets of data are obtained. One is the spectral data through the sample and the other without the sample (background or reference). Light transmission ($T$) through sample can be calculated by

$$T = I_S/I_R \qquad (11.1)$$

where $I_S$ is the sample transmission spectral scan pixel amplitude, and $I_R$ is the reference. These calculations can be performed in MATLAB® or a similar computation software. It should be noted that Equation (11.1) is valid if the image pixel value is linearly proportional to the image intensity in the range of values that cover $I_S$ to $I_R$. If the picture is saturated, or if the camera performs self calibration or self adjustment, this ratio may not be representative of the true spectrum.

The reason for using a reference signal is to avoid errors due to source variation and detector response variations. Off-the-shelf spectrometers generally use calibrated components to achieve true spectral amplitude detection, and also incorporate referencing/calibration schemes as described here by capturing data from a sample, and compare it to stored reference data.

A note of caution using flashlights as light sources: It is best to use a halogen flashlight. Some of the current flashlights in the market use white light LEDs. Although they appear white, the actual emission spectrum of many of the white LEDs has a sharp peak in blue and a broad peak in yellow (emission from a phosphor material commonly used in white LEDs). In contrast, an incandescent or halogen lamp gives a more uniform coverage of the visible spectrum than a white LED lamp.

It should be noted that this experiment if for learning purposes. If a precise spectral measurement is desired, a number of venders offer spectrometers of a variety of specifications. These units contain collection optics for more precise measurement, detector array, and detection software. Today's digital electronics and high sensitivity optical detector arrays have enabled a new generation of compact, low-cost, and high-performance spectrometers that can capture optical spectra in real time.

## EXPERIMENT 11.2: SIMPLIFIED TRANSMISSION SPECTROMETER

Purpose: To become familiar with the use of diffraction in spectroscopy
Related simulations: N/A
Materials needed:

- Incandescent or halogen flashlight
- Blank digital video disk (DVD)
- Some cardboard and tape to make an enclosure for light blocking
- Camera (e.g., a digital camera or a camera on a cell phone)
- Optional: Software to convert image data to numbers (e.g., MATLAB)

### EXPERIMENT

This experiment is a further simplification of the previous experiment. This "spectrometer" utilizes a light source (flashlight), a digital video disk (DVD) to split the light spectrum (e.g., split white light to "rainbow" colors), and a camera to detect the diffracted light.

The basic setup is shown in Figure 11.2. The setup uses an incandescent flashlight. Halogen flashlights work best. Avoid using a white LED flashlight, because the output spectrum is not as continuous in the visible spectrum as a halogen flashlight. Tape a white piece of paper on the flashlight cover glass to create a diffuse light. Place the flashlight a few centimeters above the DVD. At certain distance, the DVD will appear filled with rings of rainbow, as shown in Figure 11.3 (blue in the middle and blue at the output). This is the diffraction from LED.

Place a camera as shown in Figure 11.2 to capture the circular rainbow image shown in Figure 11.3. Next place a color filter between the camera and DVD such that portion of the circular rainbow view is blocked by the filter, as shown in the setup in Figure 11.2. If a filter is not available, a color filter can be created

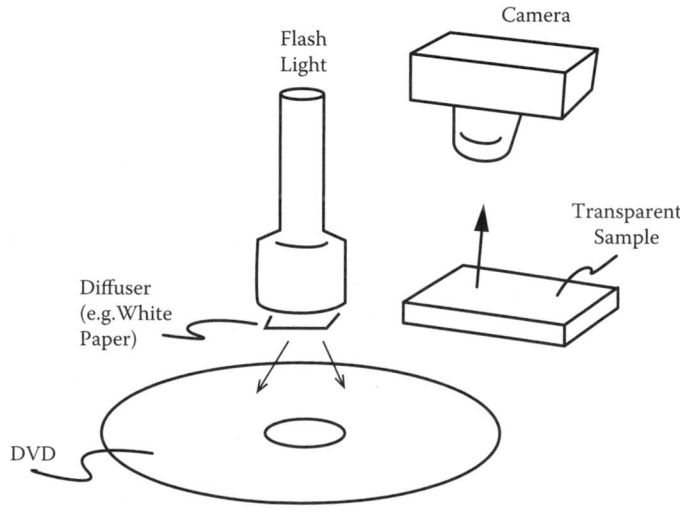

**FIGURE 11.2** Example of a further simplified spectrometer using a flashlight and diffraction from a DVD disk (read-only disk works best).

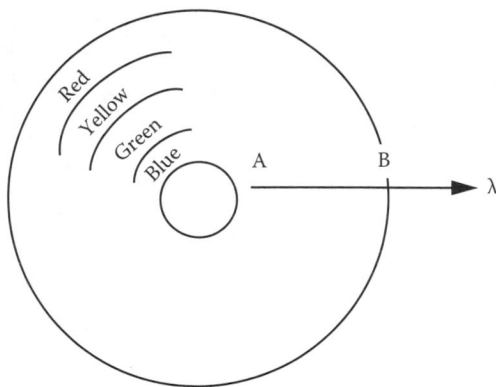

**FIGURE 11.3** Schematic of colors observed due to diffraction from the DVD using the setup shown in Figure 11.2.

by marking a clear piece of clear plastic or transparency paper with a permanent marker.

Start with a red filter, either a clear-colored plastic or glass, or create the filter by marking a clear plastic or a transparency paper by a red marker. The image portion that is covered by the red filter will transmit red light. Therefore in the portion of the image with the red filter the red rings will appear bright, and blue and green rings will be absorbed and will appear dark, as shown in Figure 11.4. Photographs of the spectra obtained by this configuration are shown in Figure 11.5a and Figure 11.5b.

Figure 11.5a shows the photograph of the diffraction from the DVD without any filter, and Figure 11.5b shows the photograph with the red filter covering the lower right side of the DVD, similar to Figure 11.4. As apparent in Figure 11.5, the red portion of the spectrum appears bright, and the blue/green portions of the spectrum appear dark.

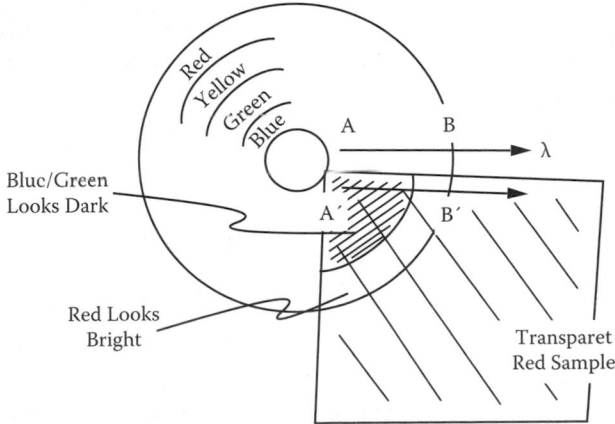

**FIGURE 11.4** Spectrum observed due to diffraction from the DVD using the setup shown Figure 11.2, with a red filter added to the setup. In the portion of the image with the red filter the red rings appear bright, and blue and green rings are absorbed and appear dark.

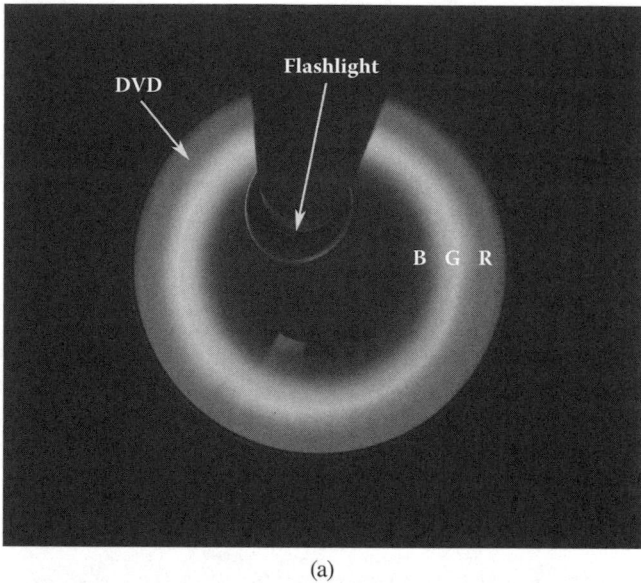

(a)

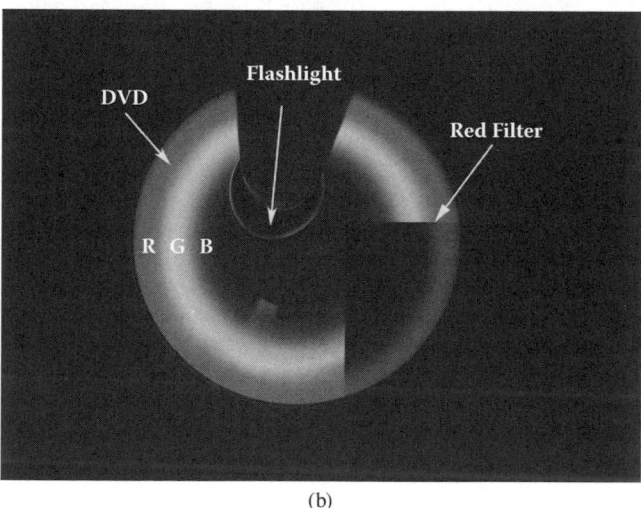

(b)

**FIGURE 11.5**  Photograph of the spectrum observed due to diffraction from the DVD using the setup shown in Figure 11.2. (a) Photograph of the diffraction from the DVD without any filter. (b) Photograph of the diffraction with a red filter covering the lower right side of the DVD. The red portion of the spectrum appears bright, and the blue/green portions of the spectrum appear dark due to red filter absorption.

To continue the experiment further, the filter transmission image can be quantified. After capturing an image with a camera similar to Figure 11.4 and the photograph shown in Figure 11.5b, read the pixel values of the image. One method to read the pixel values is to open the image file in MATLAB or any other computational software that has the ability to read image data and convert the pixel values to numbers. The experiment is preferably performed by a monochrome camera. However if this is not possible, and if the picture is an RGB image, then pixel values have to be converted to gray scale intensity values in the software. After converting to gray scale, plot the pixel values of the reference area, namely where there is no filter, as shown in Figure 11.4 from point A to point B. This will produce a plot with some variation that is dependent on the camera response. An example is shown in the top plot (solid line labeled $I_{Bkg}$) of Figure 11.6. Repeat this step in the area of the image where the filter covers portion of the DVD view (e.g., from point A' to B' in Figure 11.4) and plot the data, as shown in the top plot of Figure 11.6 (dashed line labeled $I_s$). If the two scans are taken from an area adjacent to each other as shown in Figure 11.4, then light transmission measurement can be obtained by dividing pixel data through filter with data from area without the filter (Transmission = $I_s$/$I_{BKG}$). An example is shown in the lower curve of Figure 11.6.

The y-axis of this curve (shown in the bottom of Figure 11.6) is light transmission, and the x-axis is wavelength. Note, however, to obtain exact wavelength numbers, the setup has to be wavelength calibrated. This is another optional task that can be performed by using optical filters with known transmission. One example is to use two narrow band transmission filters of known wavelength bands, such as blue and red filters, and mark where the transmission appears highest in the plotted data. This data can be used to indicate the wavelength. Note however if the apparatus is not fixed, for example, if the position of any of the components (flashlight, DVD, or camera) move, then so will the position of

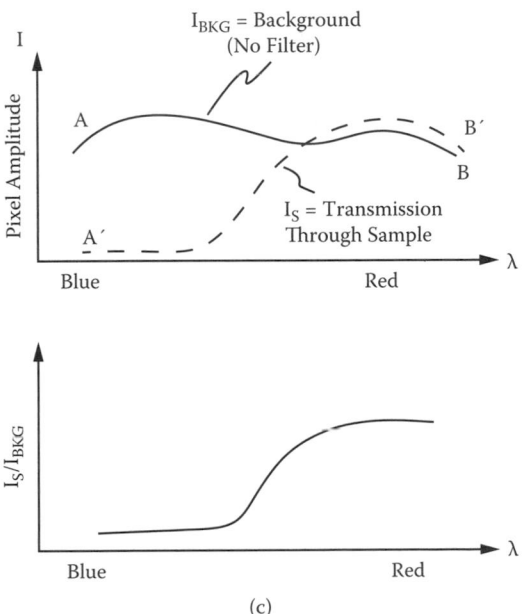

**FIGURE 11.6** Calculating filter transmission by reading the image data from the spectrum shown in Figure 11.4, and in the photograph shown in Figure 11.5b. (a) Top plot: Pixel values of the signal ($I_S$) through the filter (dashed curve) read from point A¢ to B¢ as shown in Figure 11.4. Pixel values of the background ($I_{BKG}$) without the filter (solid curve) read from point A to B as shown in Figure 11.4. (b) Bottom plot: Calculated transmission through the filter, $I_S/I_{BKG}$.

the spectrum. Therefore wavelength calibration by known filters can be achieved in a fixed setup.

### EXPERIMENT 11.3: OPTICAL BEAM PROFILER

Purpose: To familiarize with optical beam shape measurements
Related simulations: N/A
Materials needed:

- Laser pointer
- Photodiode and amplifier, or a phototransistor/resistor
- Small DC motor
- Foil, cardboard, and other materials to construct a rotating slit
- Oscilloscope

#### EXPERIMENT

In this experiment the profile of a laser beam is measured by using a rotating slit, as shown in Figure 11.7a. The apparatus consists of a rotating slit that is attached

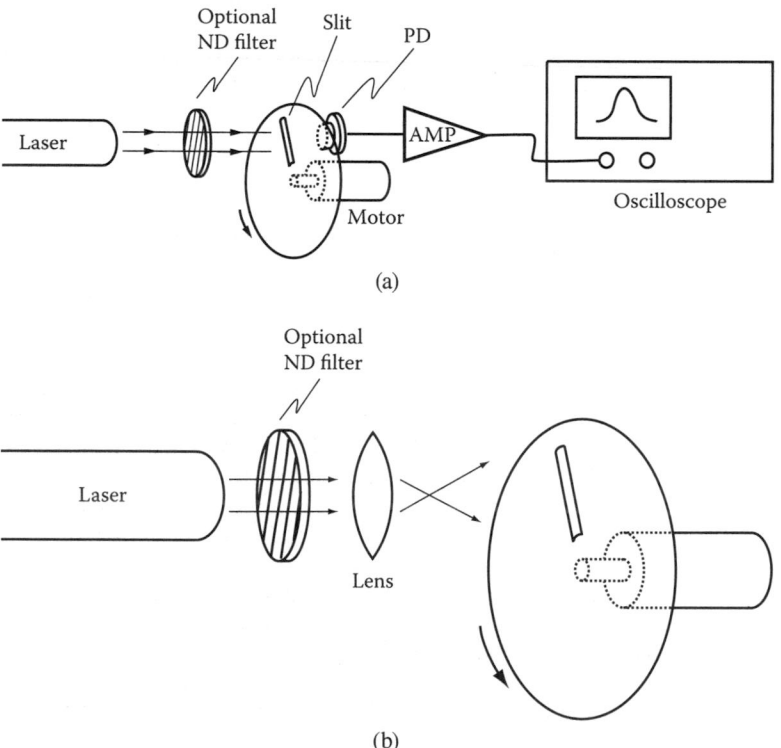

(a)

(b)

**FIGURE 11.7** (a) Laser beam profile measurement using photodiode (PD)/amplifier (shown) or a phototransistor (not shown) placed behind a rotating slit. The PD signal is sent to the oscilloscope. The signal shows the shape of the laser beam. (b) Optional lens used to expand the laser beam to make it sufficiently larger than the slit. An optional neutral density (ND) filter is used (e.g., a darkened glass) to avoid saturating the photodetector.

to a motor. On the back side of the slit a photodetector is placed. The signal from the photodetector is sent to an oscilloscope that will display the laser beam shape. The detector can be a photodiode followed by a transimpedance amplifier or a phototransistor, as described in Chapter 3.

The first step is to construct a thin slit using an aluminum foil glued or attached to a cardboard wheel or a thick paper (such as paper used for a file folder darkened with a marker). The reason for using a foil is to have a thin slit with sharp edges. Make an opening in the cardboard slightly larger than the final slit width, and then make a thin slit using the foil in the center of the wider slit. Note that the final slit width should be much narrower than the laser beam width. If this is not possible, then an alternative is to use a lens to expand the laser beam, as shown in Figure 11.7b. This, however, may distort the beam, and the true shape of the spot may be different than what is measured.

After constructing the slit, attached it to a motor as shown in Figure 11.7a. A photodiode or a phototransistor is placed behind the slit. The detector response has to be fast enough that it does not interfere with the measurement. The detector response time should be around two orders of magnitude smaller than the scan of the laser spot (e.g., if the scan displayed on the oscilloscope is 10 ms width, a 100 microsecond detector would be sufficiently fast). The oscilloscope trace shows the shape of the laser spot. If the laser output is nonsymmetric, then rotating the laser and rescanning will reveal the spot shape and size at different orientations.

This experiment demonstrates the principles of one of the laser beam scanning methods. Precision laser beam scanners are commercially available through various optical/optomechanical vendors, which result in precise spot size measurements.

### EXPERIMENT 11.4: OPTICAL FOURIER FILTERING

Purpose: To be familiar with optical filtering
Related simulations: N/A
Materials needed:

- Lenses (see Figure 11.8)
- Laser (HeNe laser or laser diode with collimating optics, similar to a laser pointer)
- Spatial filtering setup (optional)
- Object printed on a transparency
- Optomechanical hardware to mount optics
- Camera to view the image plane

Note that this experiment requires a number of optical and optomechanical components.

### EXPERIMENT

This experiment demonstrates optical spatial filtering using a 4f setup by blocking highter diffraction orders to remove periodic lines from an image. A good demonstration is to start with an image with periodic noise added to the image. An example is to generate fine periodic vertical lines (as fine as it can be printed out) using a graphics software, and add these lines to another image, such as to a

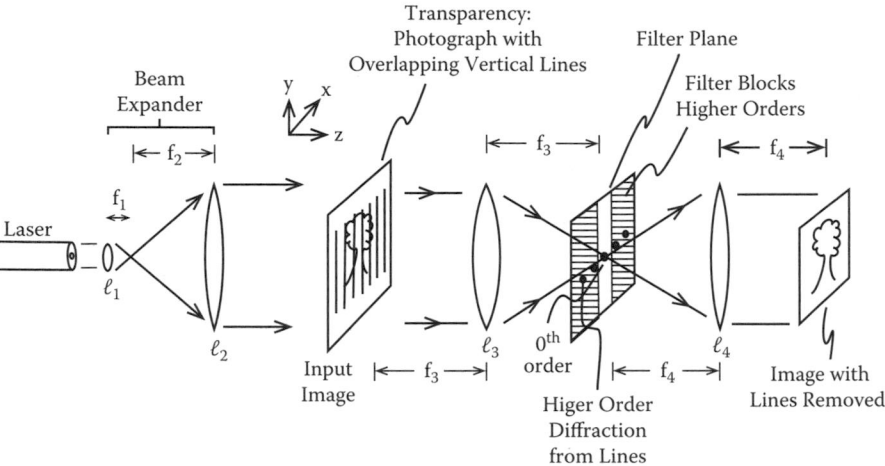

**FIGURE 11.8**  Schematic of an optical Fourier filtering setup (4f system). Laser light is expanded and collimated using lens 1 ($l_1$) and lens 2 ($l_2$). A transparency is placed one focal distance from a lens ($l_3$) that focuses light to the filter plane. A filter is inserted in the filter plane. Another lens ($l_4$) is placed one focal distance from the lens, and an image is formed at the image plane.

photograph of nature. Some of the graphics and photo editing software available off-the-shelf can be used to combine multiple images. Therefore the scene and the background noise (the periodic lines) can be combined into a single image. Print the image on a transparency sheet. If no software is available to combine the lines and the scene, then it is also possible to print the lines and the scene separately, one on a paper, the other on a transparency. Overlap them and make a copy of the overlapping image on a transparency sheet using a copy machine.

The 4f filtering apparatus is shown is in Figure 11.8. The first section is to generate an expanded laser beam that travels parallel (this is called a collimated beam) using a combination of two lenses, such as a microscope objective ($l_1$) with short focal length and a larger diameter lens ($l_2$) with longer focal length than $l_1$. In this figure, the lenses are labeled as $l_i$, and their corresponding focal lengths are labeled $f_i$. For best results this setup should have a clean uniform input light, which can be achieved by using a spatial filtering apparatus for beam expansion. However without using a spatial filter, optical filtering can still be observed with moderate success.

Continue constructing the 4f system shown in Figure 11.8. In this setup, the input image (sometime referred to as *object*) is focused at the filter plane, and reimaged back at the image plane. The filter plane (also referred to as Fourier plane) represents the spatial Fourier transform of the input image, which is the far-field diffraction pattern of the input image [2]. If the input image contains fine lines, the filter plane will have repeating spots surrounding the central spot (referred to 0th order or DC) as shown in Figure 11.8.

Initially do not insert anything in the filter plane. When the setup is complete, test it by placing the transparency at the input (object) plane, and observe the reimaging of the object in the image plane. If the lines in the input image are fine enough, then the higher-order diffraction spots resulting from the line in the input image will be visible in the filter plane. These will look like dots outside the central (0th) order, which are in the direction normal to the lines in the object. Namely, if the lines in the object are in the vertical (y) direction, the diffracted dots are in the horizontal (x) direction. If the lines are too coarse, the higher diffracted orders may be too close to each other and could not be easily filtered out with a spatial filter. The longer is the focal length of lens $l_3$, the wider is the separation of the diffracted orders. However, having a lens with very long focal length makes the setup very large and unmanageable.

When the apparatus is complete, insert a slit in the filter plane such that higher-order spots are blocked. Observe that the lines in the image disappear. Remove the slit from the filter plane, and the lines reappear, thus illustrating the filtering capabilities of this optical setup. Also observe that some parts of the image, particularly vertical edges, will look blurred because of the spatial filtering effect.

An alternative to using a slit in the filter plane is to use a filter that only blocks the higher order diffraction spots (e.g., two horizontal rectangles separated from each other in the x-direction), but the rest of the light passes through the filter. Therefore the original image quality will be less affected than using the slit shown in Figure 11.8, but the vertical lines will still be filtered out.

This setup is best achieved using a camera to view the image plane. The camera can be placed either behind the image plane using a thin paper in the image plane or placed on the side to view the image plane from the front at an angle.

### OTHER USES OF 4F OPTICAL FILTERING SYSTEM

The 4f system shown in Figure 11.8 is also called an optical coherent processing system [2]. It is often used as an optical correlator, where the output image, $I_{out}$, is the correlation of the input image and a reference image, namely,

$$I_{out} = I_{in} \otimes I_{ref} \tag{11.2}$$

where $I_{in}$ is the input image, and $I_{ref}$ is a reference image. The filter used in the filter plane is the Fourier transform of the reference image $I_{ref}$. Because the Fourier transform of an image is in general a complex function, this filter has to be represented by a combination of amplitude and phase. Conventional photographic film only records amplitude data and therefore has a limited functionality as a filter material. To achieve phase and amplitude representation in the filter plane, earlier work in this area was performed by generating the complex (amplitude and phase) filter function via holography, by adding a reference configuration to the optical processing setup [3]. The holographic techniques, however, resulted in fixed optical filter that was not easily upgradable. Later work showed that it is possible to generate real-time (video rate) upgradable filters using liquid crystal light valve spatial light modulators (SLMs) [4]. The technique combined the amplitude and polarization modulation scheme of the SLM to enable three state phase and amplitude representation. Therefore a 4f system can be a valuable tool for high-speed correlation, particularly when the input image and the camera can be updated at high speed.

Optical filtering and correlation is often used in setups where extreme high-speed filtering is required. There are several bottlenecks in the system that limits the time to achieve optical filtering. These are (1) time to update the input image, (2) time to update the filter, and (3) the camera capture speed. Many of the modern high-speed spatial light modulators and high-speed cameras enable extremely high optical processing.

## MAKING THE SYSTEM COMPACT

To obtain a diffraction pattern in the filter plane with the diffracted orders widely separated to enable spatial filtering, the required lens focal length may be very long. Some advanced techniques exist to construct the 4f setup in a compact form. One of them is to utilize a combination of a positive and negative lens instead of a single positive lens. The purpose of adding a negative lens is to increase the diffracted spot size, while keeping the length of the system lower than using a single positive lens. This is achieved by using a negative lens of a shorter focal length than the positive lens, and placing it between the positive lens and the filter plane. The focal points of the positive and negative lenses should overlap at the filter plane. This will result in a much larger diffraction spot than without using the negative lens.

## EXPERIMENT 11.5: POLARIZED RAYLEIGH SCATTERING WITH MILK AND WATER

Purpose: To familiarize with scattering and polarization
Related chapter/section: 5–Polarization
Related simulations: N/A
Materials needed:

- Glass container with flat sidewalls
- Halogen/incandescent (non-LED) flashlight (preferably dim)
- Milk and water
- Two pieces of polarizers (plastic polarizer sheets work well)

Caution: Do not stare directly into a bright flashlight. Use sunglasses if necessary, and do not view directly into the light beam.

### EXPERIMENT

This is a continuation of the Experiment 4.3 from Chapter 4. That experiment was to observe the color change due to scattering. This experiment is to observe polarization of scattered light.

When light encounters suspended particles that are small compared to the wavelength of the light, scattering is polarization in a particular direction. This is called Rayleigh scattering, and it is the reason why the sky is polarized, as described in Chapter 5 and illustrated in Experiment 5.5. The aim of this experiment is to mimic a small particle scattering phenomenon, which can be achieved by adding a small amount of milk to water.

The experimental setup is shown in Figure 11.9a. Fill a glass container with water. Use an incandescent or halogen flashlight (not an LED flashlight) as a light source. Add a small amount of milk to the water. This will make the water murky. The amount of milk should be such that the filament of the flashlight should be barely visible when viewing light transmission through the milk–water mixture if viewed from the opposite side of the glass container. Be sure not to use a very bright flashlight, and use dark neutral density (gray) sunglasses if necessary. Do not stare directly into the beam. Instead view at an angle.

Place a glass plate on top of the container holding the milk–water mixture and place two polarizers on top as shown in Figure 11.9a (typically plastic polarizer

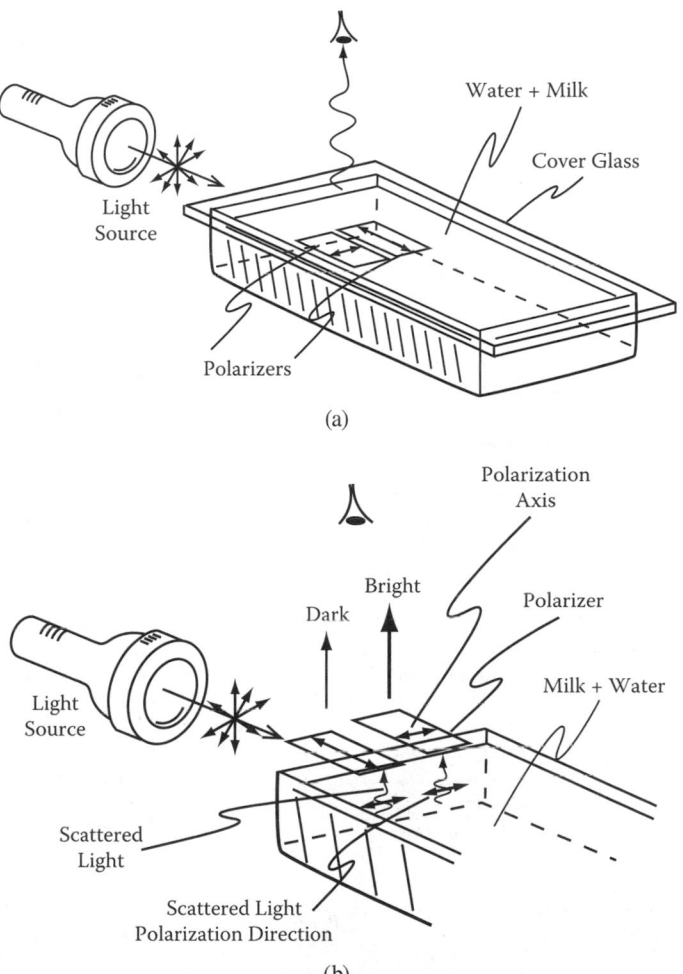

**FIGURE 11.9** (a) Experimental setup to observe polarization orientation of light scattered from a milk–water mixture. Two polarizers are placed on a cover glass. One polarizer has the polarization axis in the direction of light propagation, and the other is perpendicular to the light propagation. (b) Details of the setup. If the polarizer axis is parallel to the incident light, the polarizer looks dark to the observer. If the polarizer axis is perpendicular to the incident light, the polarizer looks bright to the observer because it is parallel to the scattered light polarization orientation.

are available from vendors such as Edmund Optics work well). The polarizers are oriented such that one has the polarization axis in the direction of light propagation, and the other is in direction perpendicular to the light propagation, as shown in Figure 11.9a and detailed in Figure 11.9b. When viewing light from the top, scattered light through the polarizer oriented with a polarization direction parallel to the light propagation direction will be less (looks darker) than scattered light from the polarizer with a polarization direction perpendicular to the direction of light propagation. This is because the scattered light is primarily oriented in the direction perpendicular to light propagation, as detailed in Figure 11.9b.

Photographs of this experiment are shown in Figure 11.10a and Figure 11.10b, taken from top, at the viewing angle as indicated in Figure 11.9a. In Figure 11.10a two polarizers are placed next to each other: one with the polarization angle oriented parallel to the light propagation, and the other perpendicular to the light propagation direction. The polarizer axis is indicated with arrows marked on the polarizers. As apparent in Figure 11.10a, the area with the polarizer oriented parallel to light propagation appears darker than the polarizer oriented in the perpendicular direction (this is more apparent close to the center of the optical axis,

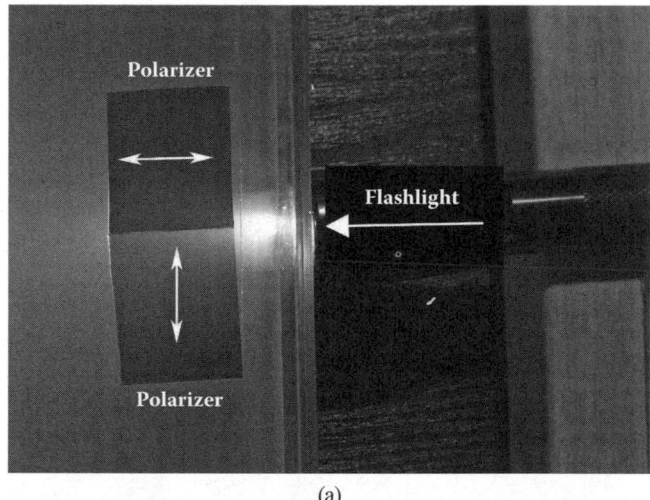

(a)

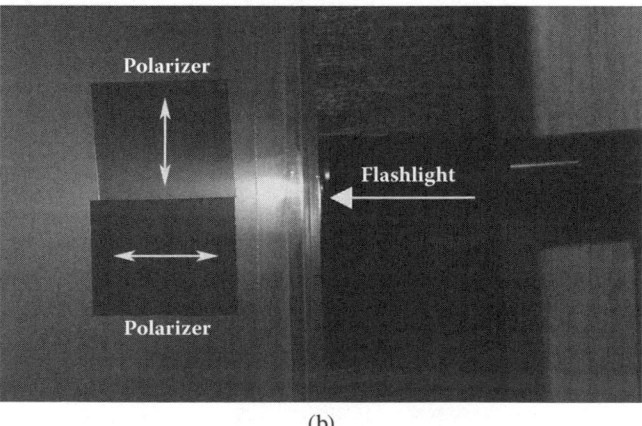

(b)

**FIGURE 11.10** Photographs of the experimental setup shown in Figure 11.9 from the position of the observer. The polarizer orientations are reversed in (a) and (b). In both pictures (a) and (b), if the polarizer axis is parallel to the incident light, the polarizer looks dark. If the polarizer axis is perpendicular to the incident light, the polarizer looks bright.

where light is brightest). The observed difference in brightness is due to the fact that the polarization of light scattered from the milk–water mixture toward the camera is perpendicular to light propagation direction.

The same experiment is repeated by rotating the orientation of the polarizers, and a photograph is taken as shown in Figure 11.10b. The same effect is observed, namely, light through the polarizer oriented parallel to the light propagation direction appears darker than the polarizer oriented perpendicular to the light propagation direction.

Compare the findings of this experiment to Experiment 5.5.

## REFERENCES

1. Pedrotti, F. L., and L. S. Pedrotti. 1987. *Introduction to Optics*. Englewood Cliffs, NJ: Prentice-Hall.
2. Goodman, J. W. 2004. *Introduction to Fourier Optics*. 3rd ed. Greenwood Village, CO: Roberts & Company.
3. Collier, R. J., C. B. Burckhardt, and L. H. Lin. 1971. *Optical Holography*. San Diego, CA: Harcourt Brace Jovanovich.
4. Chao, T. H., A. Yacoubian, B. Lau, E. R. Hegblom, and W. J. Miceli. 1994. "Optical Wavelet Processor Using Cascaded Liquid Crystal Television Spatial Light Modulators." *Proceedings of SPIE on Photonics for Processors, Neural Networks, and Memories II*, Vol. 2297, 23–32.

# 12 Advanced Topics

This chapter describes some of the advanced topics not covered throughout this book. Because of the diverse applications of optical sciences, it is simply not possible to cover all of them in one chapter, but the reader is encouraged to search other topics of interest. Good places to start are trade journals and magazines published by optics organizations (e.g., The Optical Society, www.osa.org; SPIE, www.spie.org; and IEEE Photonics Society, www.photonicssociety.org), as well as symposia organized by these societies.

*Fluorescence.* Some optical material, when illuminated with a certain wavelength (e.g., ultraviolet light), emit light at different wavelengths (e.g., blue or green). Light emitted often has a lifetime, where emitted intensity decays over time, and the decay time can be used to characterize the material. Many biological materials exhibit such behavior. Applications of fluorescent materials and testing include applications in cell biology, testing for food contaminants, making security features, and building toys (e.g., stars that glow in the dark).

*Coherence.* A distinguishing feature of laser light is its long coherence length compared to natural lighting [1],[2]. Coherent light is when light waves at different spatial positions move in unison, and therefore can be made to interfere. When light is coherent in the direction of light travel, it is called temporally coherent. When light is coherent in the direction of normal to light travel, it is called spatially coherent. The distance in which it is coherent is called coherence length. Many applications such as holography and interferometry were enabled or became widely used due to advancement of coherent light sources such as lasers.

*Interferometry.* Interferometry is enabled by interfering two or more coherent light waves. When two light waves interfere, they generate fringes that can be detected by photodetectors or optical recording media. When one of the light paths is altered even slightly, then fringes move. Very small changes (much shorter than the optical wavelength) in distance produce detectable changes in the fringes. For example, a white light interferometric microscope can detect surface topography on the order of angstroms while using visible light with wavelength of hundreds of nanometers.

*Fourier optics.* As discussed earlier, optical diffraction has a form of Fourier transform (FT) [3]. Using this property of diffraction, optical systems can be configured so that they can be used for various filtering applications. For example, a 4f optical system can be used to generate the FT of a test pattern, use a spatial filter in the Fourier plane, and inverse Fourier transform to an image plane, thus enabling a highly parallel and very rapidly correlation calculations, with the only limitation (beside the speed of light) being the image input and output from electronic systems.

*Holography.* Holography is the science of recording phase and amplitude in an optical media, recording of three-dimensional images in two-dimensional media, or recording of very large amounts of data in a very small volume [4,5]. Holograms are generated by interfering coherent light. When two coherent light waves interfere, they produce fringes that are recorded by an optical media (such as silver halide photographic film or a dichromated gelatin film). If one of the light waves is altered, such as it is scattered from an object, then the recorded fringes are spatially modulated, and this modulation contains information both on amplitude and phase, unlike a standard photograph, which only records light amplitude at each position on the film. The recorded hologram can then be played back by another light source, and light diffracted from the fringes produces three-dimensional objects. There are many areas of holography, such as display holography, volume holography, holographic data recording, and real-time holography. Examples include display holograms used widely on credit cards, which are embossed white-light holograms produced from a holographic master.

*Nonlinear optics.* One area of nonlinear optics is harmonic generation. As discussed in Chapter 5, when light propagates in a medium, the wavelength does not change. This is true for most linear materials. There are cases, however, such as when using a nonlinear medium, where light wavelength does change if excited with high-intensity light. For example, when a nonlinear medium is exposed to a high-intensity laser pulse (e.g., infrared wavelength at 1064 nm), it is possible to generate a second harmonic (e.g., generate 532 nm). This phenomenon is called harmonic generation. It is used in a variety of applications, such as generation of second harmonic (green) light from an infrared mode-lock laser [6–8].

Other areas of nonlinear optics include saturable absorbers, where light absorption is dependent on the light intensity. The more intense the light, the higher the absorption. Example of use of saturable absorbers is in mode-locked lasers, enabling high-power, short-pulse lasers in a compact format.

*Linear electro-optics.* This is a branch of optics where optical materials are used to modulate the intensity of an incoming light using an electrical signal. In electro-optic materials, voltage applied across the material causes small changes in the refractive index of the material in a particular direction. When these materials are incorporated in an interferometer, such as adding them to one path of an interferometer, and if an electrical signal is applied, then this causes the fringes to move, subsequently changing the intensity of light at the output of the interferometer. Such devices are called optical modulators. Some of these devices utilize planar waveguides, where the electro-optic material is incorporated in the waveguide devices. Examples of electro-optic materials include $LiNbO_3$ crystals [6] and electro-optic polymers.

*Spectroscopy.* Spectroscopy is the field of optics that involves characterizing an incident broad spectral light by producing data that indicates intensity variation with respect to optical wavelength. Examples of such instruments is the Czerny-Turner spectrometer, where a broad incident light (e.g., white light covering several hundred nanometers) entering an input slit is collimated by a concave mirror, diffracted by a diffraction grating, and refocused to an exit slit by another concave lens. The diffraction grating diffracts light, and only narrow wavelength band (e.g., 0.1 nm) light passing through the exit slit is then detected by a photodetector. When the grating is rotated, light coming out of the exit slit changes color (wavelength scan). Rotating the grating while recording the photodetector signal will yield the light spectrum. Many of the modern-day spectrometers utilize linear array detectors that can produce optical spectrum in real time, without having the need to use moving parts. Some of the subfields of spectroscopy include imaging spectrometry and hyperspectral imaging, where a broad spectral image is separated to narrow spectral bands. Examples of applications of spectroscopy include material characterization, characterizing light sources, and crop monitoring.

*Optical shop testing.* Optical shop testing [9] is the field where optical components such as lenses and mirrors are tested and characterized for flatness and aberrations. Examples of techniques that are used in optical shop testing are interferometry, knife-edge testing, and Ronchi testing and Moire techniques. These techniques produce data and images that represent various types of aberrations and surface quality. For example, in an interferometric setup if an optical component is expected to produce straight fringes, a distorted fringe may indicate incorrect polished area, such as a dip. Optical shop testing methods are commonly used both during fabrication of optical components to achieve high-quality surfaces and to qualify optical components after fabrication.

*Optical storage.* A number of optical storage devices have been proposed and researched for using various optical techniques to store data. Compact disk (CD) players are an example of optical storage that found mainstream success, where data is stored in the form of pits in a plastic layer. Another storage technology is the magnetooptical drive, which uses magnetooptical material to record data. A high-intensity laser heats a spot on the disk while an electromagnet records the data, and reading is achieved by reading the reflection with a low-intensity laser and polarization components. Other technologies that have been researched and are being investigated include holographic storage in crystalline and organic materials.

*Optical sensing.* Optical sensing is a broad field where optical components and materials are used for various sensing applications. These systems utilize optical effects such as interferometry, metal optics (plasmonics), diffraction effects, spectrometry, polarization and birefringence, and guided wave optics. One discipline of optical sensing are fiber sensors [10,11], where optical fibers are used in a number of applications such as for environmental sensing, structural health monitoring (e.g., fibers incorporated in bridges to monitor vibrations and aging), and temperature sensing. Fibers and other sensors are increasingly becoming essential parts of medical applications for various imaging and health monitoring.

*Photonics crystals.* Photonics crystals allow fabrication of planar (and three-dimensional) optical devices in compact format. They are somewhat analogous to diffraction gratings, using periodic structures and with intentionally introduced defects. However, unlike diffraction gratings, such as volume holographic gratings, that can have very small changes in refractive index within one grating period (e.g., $\Delta n = 0.005$), photonic crystals can have very large changes in refractive index within the periodic structure, such as $\Delta n = 0.5$ or higher. This large refractive index contrast allows for very sharp (e.g., 90 degree) turns in light structure, and therefore light guiding in a very compact space. Some of the challenges are mass production of such devices because of the very small features, and in- and out-coupling of light to other devices.

*Nonimaging optics.* Nonimaging optics [12,13], as the name suggests, involves optics that are not concerned with imaging but rather efficient collection and transfer of light. Applications include solar light collectors and artificial illumination. The basic starting principles for nonimaging optics are the same, namely, using reflection and refraction principles for ray tracing. However, since the concern is not imaging, some of the imaging parameters, such as resolution and aberrations, are not investigated. Instead, design methodologies aimed at maximizing light transfer are used, as described in nonimaging optics texts [12,13].

Other disciplines in optics include *plasmonics and metal optics* [1], which covers optics of conductive materials; *super resolution imaging,* which include topics to overcome diffraction limits of optics; *structured illumination*; various disciplines of *microscopy, diffractometry, optical computing* [14], and optical neural networks; *nanomaterials* used in optics; and *negative refractive index* materials. *Optical coherence tomography* (OCT) is used to obtain cross-sectional view of materials, a non-invasive method of obtaining cross-sectional view of the retina, and obtaining cross-sectional images of arteries using fiber-optic probe-based OCT. Other topics also include optical *metrology, optical coating* and thin film design, *optical phase conjugation* [6], and *adaptive optics* used to correct distortion in telescopes due to atmospheric phenomenon in real time.

## REFERENCES

1. Born, M., and E. Wolf. 1999. *Principles of Optics: Electromagnetic Theory of Propagation, Interference and Diffraction of Light.* 7th ed. New York: Pergamon Press.
2. Goodman, J. W. 2000. *Statistical Optics.* New York: Wiley Interscience.
3. Goodman, J. W. 2004. *Introduction to Fourier Optics.* 3rd ed. Greenwood Village, CO: Roberts & Company.
4. Collier, R. J., C. B. Burckhardt, and L. H. Lin. 1971. *Optical Holography.* San Diego, CA: Harcourt Brace Jovanovich.
5. Cathey, W. T. 1974. *Optical Information Processing and Holography.* New York: John Wiley & Sons.
6. Yariv, A., and P. Yeh. 2002. *Optical Waves in Crystals: Propagation and Control of Laser Radiation.* New York: Wiley Interscience.
7. Boyd, R. W. 2008. *Nonlinear Optics.* 3rd ed. Orlando, FL: Academic Press.
8. Binh, L. N., and D. V. Liet. 2012. *Nonlinear Optical Systems: Principles, Phenomena, and Advanced Signal Processing (Optics and Photonics).* Boca Raton, FL: CRC Press.
9. Malacara, D. 2007. *Optical Shop Testing.* 3rd ed. New York: Wiley Interscience.
10. Dakin, J., and B. Culshaw. 1988. *Optical Fiber Sensors: Principles and Components.* Vol. 1. Boston: Artech House.
11. Dakin, J., and B. Culshaw. 1988. *Optical Fiber Sensors: Principles and Components.* Vol. 2. Boston: Artech House.
12. Chaves, J. 2008. *Introduction to Nonimaging Optics.* Boca Raton, FL: CRC Press.
13. Winston, R., J. C. Miñano, and P. Benitez. 2005. *Nonimaging Optics.* Burlington, MA: Elsevier Academic Press.
14. Caulfield, H. J., and G. Gheen. 1989. *Selected Papers on Optical Computing.* Washington: SPIE Optical Engineering Press.

# Appendix A: Simulations

Simulations will have one or both of the following formats.

- Script format, where the function/script is run by hitting the run button (e.g., in MATLAB® 2007, Debug → Run (F5)).
- Some of the simulations will also have the graphic user interface (GUI) version. In this case, when you run the function, GUI appears with the corresponding controls and display. These MATLAB files were tested on MATLAB 2007b. Earlier versions may not work properly.

A word of caution: These simulations are to help the reader visualize and understand optical concepts. They are not guaranteed to be error free, and they are not intended to be used as ray tracing or optical design software. These simulations are not replacements for commercially available software. Additionally, these simulations are not substitutes for professional help for designing optical components and systems.

## SIMULATION A.1: BLACKBODY RADIATION

Simulation file: Blackbody_radiation.m
Purpose: To become familiar with emission spectra of hot objects
Related chapter: 2–Light Sources

### INSTRUCTIONS

1. In the upper menu: Debug → Run. (Or use F5. In MATLAB 2007b or later, use "Run" arrow icon on the menu.)
2. The program will ask: "Input temperature in deg. Kelvin:"
3. Enter temperature of the hot object (e.g., 2800 for incandescent light bulb).

A normalized emission spectra will be plotted. An example of an output plot for temperature of 3000 K is shown in Figure A.1.

### NOTES

- Temperature range for this simulation: 20 K to 10000 K.

## SIMULATION A.2: REFRACTION

Simulation file: Refraction.m
Purpose: To become familiar with principle of refraction
Related chapter: 4–Manipulation of Light

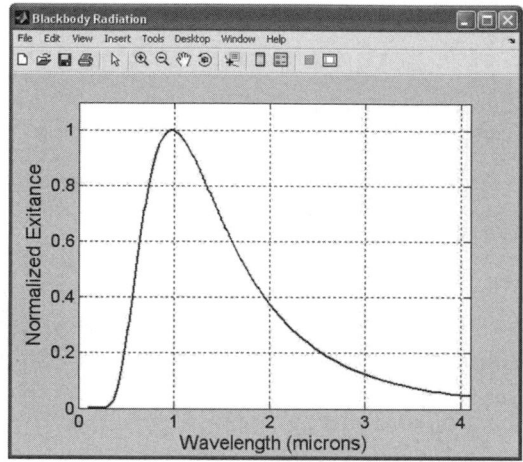

**FIGURE A.1** Screen capture of the plot generated by running the MATLAB simulation "Blackbody_radiation.m." For this example, object temperature was 3000 K.

### INSTRUCTIONS

1. In the upper menu: Debug → Run. (Or use F5. In MATLAB 2007b or later, use "Run" arrow icon on the menu.)
2. The program will ask: "Enter Refractive index in medium 1 (n1):" Enter refractive index of the first medium.
3. The program will ask: "Enter Refractive index in medium 2 (n2):" Enter refractive index of the second medium.
4. The program will ask: "Input angle (deg.):" Enter angle of incidence in the first medium.
5. The output will be plotted and the output angle will be given at the input prompt.

### Example 1

Enter Refractive index in medium 1 (n1): 1
Enter Refractive index in medium 2 (n2): 1.5
Input angle (deg.): 40
Output angle = 25.374 (deg.)

### Example 2 (Input Angle Is Larger than Critical Angle)

Enter Refractive index in medium 1 (n1): 1.5
Enter Refractive index in medium 2 (n2): 1
Input angle (deg.): 50
Output angle = –50 (deg.)

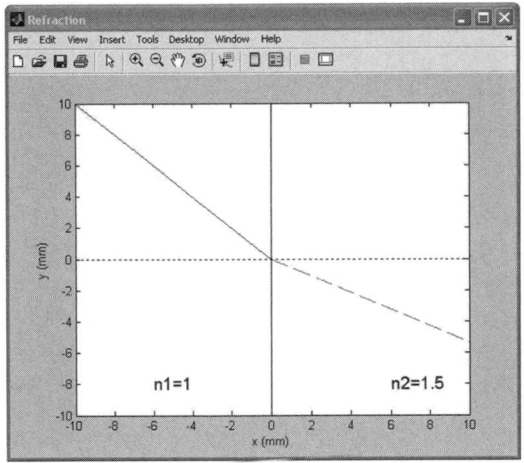

**FIGURE A.2**   Screen capture of the plot generated by running the MATLAB simulation "Refractio.m." For this example, n1 = 1, n2 = 1.5, Input angle = 45 deg., and the resulting output angle = 28.13.

## NOTES

- When light is refracted, a portion of it is reflected from the interface. This reflection is not shown for clarity.
- If the input angle is larger than the critical angle (as specified in Chapter 8), then instead of refraction, you will see a total internal reflection.
- Valid for input angle range 0 to ±90.
- Example is shown in Figure A.2.

## SIMULATION A.3: REFRACTION (GUI)

Simulation file: Refraction_GUI.m
Purpose: To become familiar with the principle of refraction. Same as Refraction.m, however uses graphic user interface (GUI).
Related chapter: 4–Manipulation of Light

### INSTRUCTIONS

1. In the upper menu: Debug → Run. (Or use F5. In MATLAB 2007b or later, use "Run" arrow icon on the menu.)
2. Change GUI inputs and observe light rays. To change $O_1$, either use sliders or enter number in the box, then move the mouse outside the box and click. To change $n_1$ or $n_2$, enter number in the coresponding box, and click the mouse outside the box.

Input and output angles are shown at the MATLAB prompt.

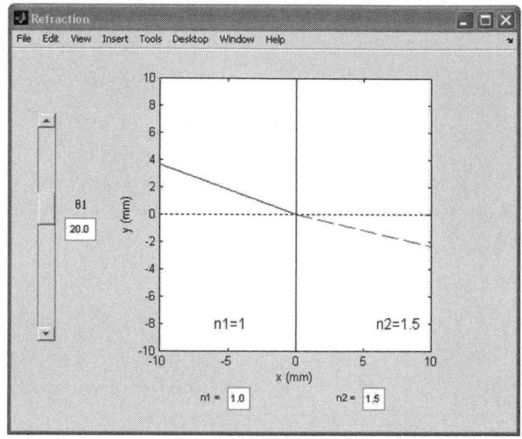

**FIGURE A.3** Screen capture of the plot generated by running the MATLAB simulation "Refractio_GUI.m."

## Notes

- When light is refracted, a portion of it is reflected from the interface. This reflection is not shown for clarity.
- If the input angle is larger than the critical angle (as specified in Chapter 8), then instead of refraction, you will see a total internal reflection.
- Valid for input angle range 0 to ±90.
- An example is shown in Figure A.3.

## SIMULATION A.4: DIFFRACTION (GUI)

Simulation file: Diffraction_GUI.m
Purpose: To become familiar with principle of diffraction
Related chapter: 4–Manipulation of Light

### Instructions

1. In the upper menu: Debug → Run. (Or use F5. In MATLAB 2007b or later, use "Run" arrow icon on the menu.)
2. Change GUI inputs and observe light diffraction. Either use sliders or enter number in the box, then move the mouse outside the box and click.

### Notes

- The lower plot is the input pattern transmission (aperture function), and the upper plot is the normalized diffracted intensity ($|U(x_o)|^2$ of Equation (B.4) in Appendix B. This simulation is valid when the input to the aperture is a plane wave, such as a collimated laser beam.

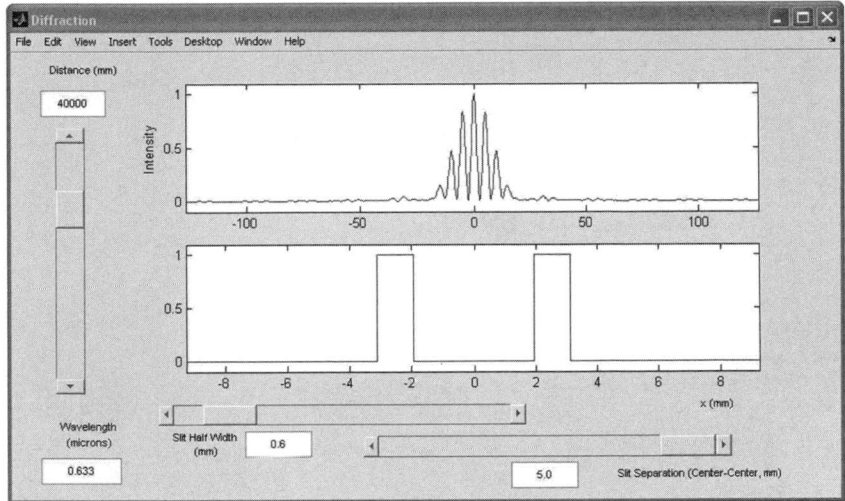

**FIGURE A.4** Screen capture of the plot generated by running the MATLAB simulation "Diffraction_GUI.m." Slit parameters are set such that the input pattern is a double slit. (Slit width = 0.6 mm; slit separation = 5 mm center to center.)

- If the slit separation is large enough, then a double slit diffraction is observed (see Figure A.4). If the slit separation is small such that input plot shows a single slit pattern, then the output will be diffraction from a single slit (see Figure A.5).
- The calculations are valid when in Fresnel or Fraunhofer zones, as discussed in Appendix B. If these conditions are not met, the following will be stated in MATLAB.
    - Not in Fresnel or Fraunhofer zone.
    - Try larger z.

- There are some instances, depending on the input parameters, the plotted diffraction pattern may not be indicative of the true diffraction pattern. This could happen due to round-off errors and other computing errors.

## SIMULATION A.5: POLARIZATION

Purpose: To become familiar with polarizers, quarter- and half-wave plates.
Related chapter: 5–Polarization
Simulation file: Polarization_GUI.m

### INSTRUCTIONS

1. In the upper menu: Debug → Run. (Or use F5. In MATLAB 2007b or later, use "Run" arrow icon on the menu.)
2. Change GUI inputs and observe light polarization effects. Either use sliders or enter number in the box, then move the mouse outside the box and click.

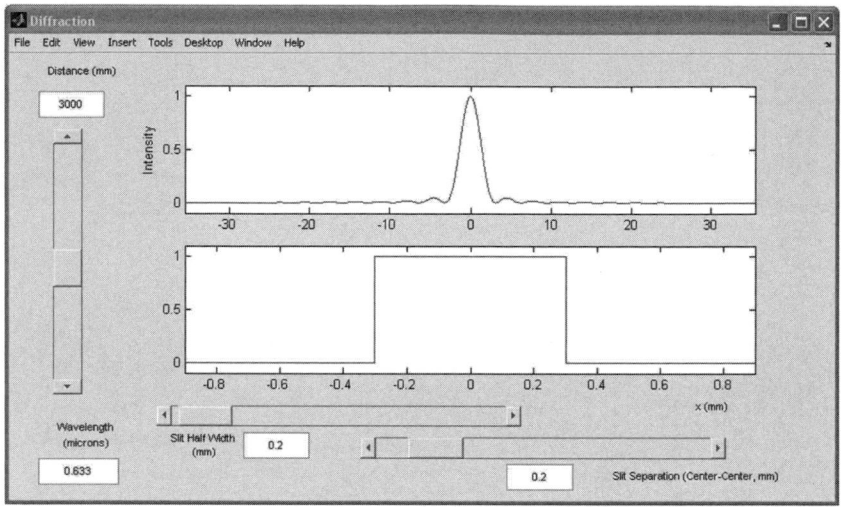

**FIGURE A.5** Screen capture of the plot generated by running the MATLAB simulation "Diffraction_GUI.m." Slit parameters are set such that the input pattern is a single slit. (Slit width = 0.2 mm; slit separation = 0.2 mm center to center.)

Use of quarter- and half-wave plates, and the output polarization can be visualized with this simulation. This simulation has two options: use only two components or use three components.

Option 1: Two components—In the Component Selection box, if the top row is selected, then only one component is used after polarizer P1. The component is either a half-wave plate ($\lambda/2$), a quarter-wave plate ($\lambda/4$), or another polarizer (P2). The first polarizer angle is selected by the slider bar labeled P1 Angle (or value entered at the box below). If a wave plate is selected in the Component Selection box, then its orientation is selected by the Wave Angle slider bar (or the box below). An example of an input polarizer (P1) followed by a quarter-wave plate is shown in Figure A.6.

Option 2: Three components—In the Component Selection box, if the bottom row is selected, then two components are used after polarizer P1. The middle component is either a half-wave plate ($\lambda/2$) or a quarter-wave plate ($\lambda/4$). Following the wave plate, a second polarizer (sometimes called analyzer) is used, P2. Component angles are selected by the sliders (or the box below). Since there is a polarizer at the output (P2), then the polarization will always be linear, orientated at the direction of P2. However, the relative transmitted intensity will be different, which is indicated by the bar to the right ($I_{Rel}$). In calculating the transmitted intensity, $I_{Rel}$ surface reflections, and absorption by the polarizers and wave plates are ignored, and therefore this is a relative measurement. An example of an input polarizer (P1) followed by a quarter wave plate and another polarizer (P2) is shown in Figure A.7.

In both Option 1 and 2, the input polarization angle is selected by "P1 Angle."

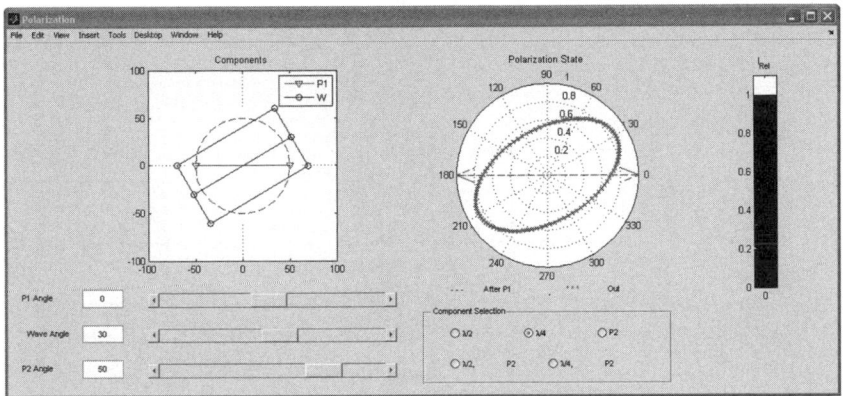

**FIGURE A.6** Screen capture of the plot generated by running the MATLAB simulation "Polarization_GUI.m." The example shows an input polarizer (P1) followed by a quarter-wave plate.

The polar plot in the middle shows the input polarization after the first polarizer (P1), in blue dotted line, and the right is polarization orientation at the output. The plot to the left shows the component orientation, circles labeled P1 and P2 for the polarizers; and the square labeled W is the orientation of the wave plate.

## SIMULATION A.6: IMAGING WITH A SINGLE LENS (GUI)

Simulation file: SingleLens_GUI.m
Purpose: To become familiar with image generation using a single lens
Related chapter: 6–Geometrical Optics

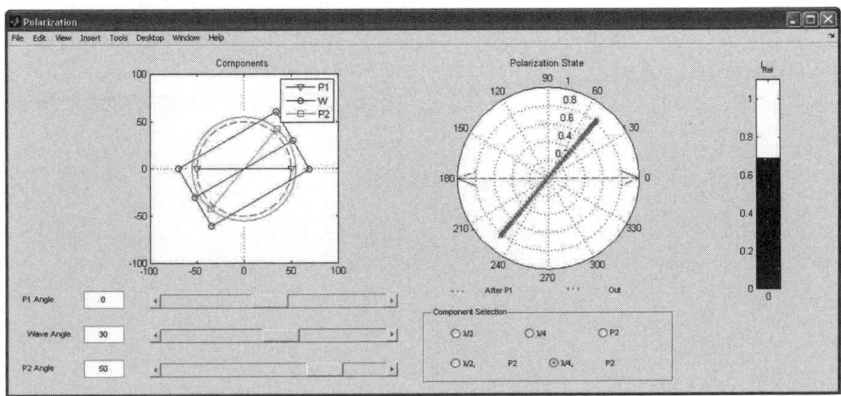

**FIGURE A.7** Screen capture of the plot generated by running the MATLAB simulation "Polarization_GUI.m." The example shows an input polarizer (P1) followed by a quarter-wave plate and another polarizer (P2).

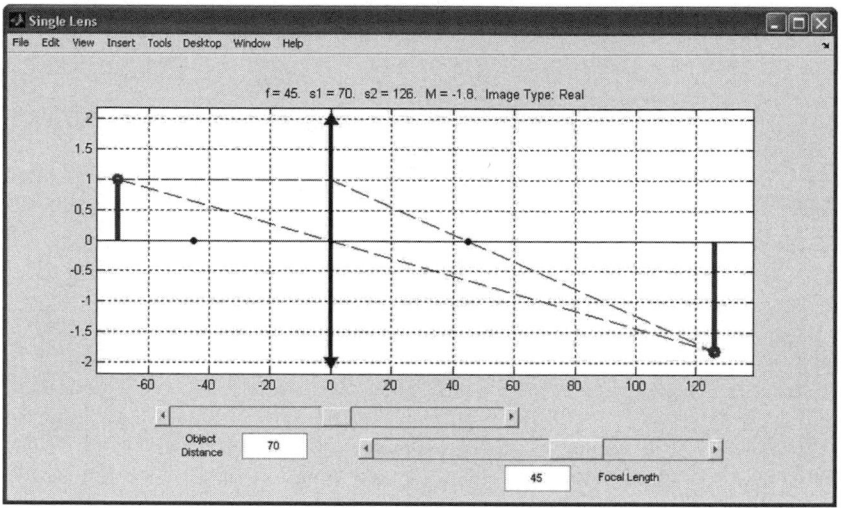

**FIGURE A.8**   Screen capture of the plot generated by running the MATLAB simulation "SingleLens_GUI.m." The example shows a positive lens with the focal length producing a real image.

## INSTRUCTIONS

1. In the upper menu: Debug → Run. (Or use F5. In MATLAB 2007b or later, use "Run" arrow icon on the menu.)
2. Change GUI inputs and observe image location. Either use sliders or enter number in the box, then move the mouse outside the box and click.

## NOTES

- When the image is real, it is shown in solid blue line (see Figure A.8). When the image is virtual, it is shown in red dashed line (Figure A.9).
- When the object is placed right at the focus, then the image is at infinity, and no calculations will be performed. The following will show in MATLAB prompt: "Nearly collimated. Change object distance or focal length."

## SIMULATION A.7: RESOLUTION (GUI)

Simulation file: Resolution_GUI.m
Purpose: To become familiar with image degradation due to resolution limit
Related chapter: 7–Imaging Systems

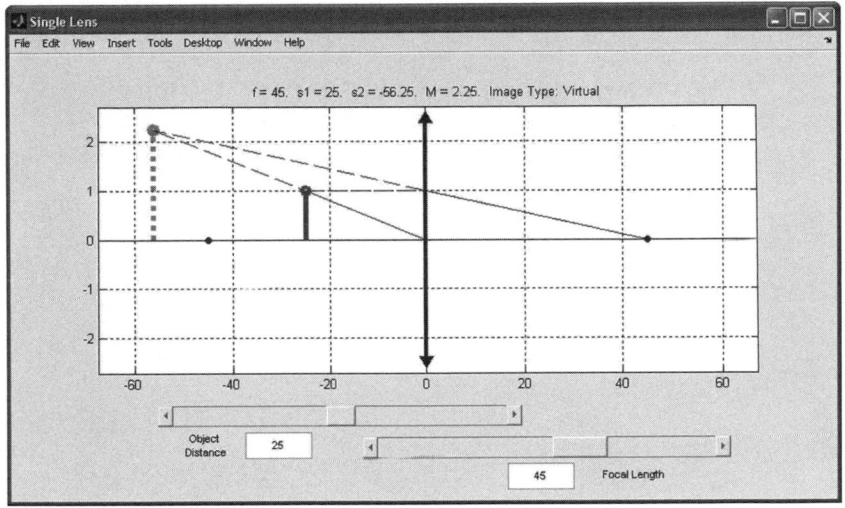

**FIGURE A.9**   Screen capture of the plot generated by running the MATLAB simulation "SingleLens_GUI.m." The example shows a positive lens producing a virtual image.

## INSTRUCTIONS

1. In the upper menu: Debug → Run. (Or use F5. In MATLAB 2007b or later, use "Run" arrow icon on the menu.)
2. Change GUI inputs and observe image location. Either use sliders or enter number in the box, then move the mouse outside the box and click.

## NOTES

- This is a one-dimensional simulation.
- The top image represents a line pattern with varying pitch (this is the input image).
- The middle image (labeled "Filtered") is the filtered image with.
- The filter cutoff spatial frequency is w, in lines per millimeter.
- w can either be entered in the box at the bottom of the GUI or can be changed using the slider.
- The lower plot shows the *input* and *filtered* amplitudes, in red *x* markings and in blue solid lines, respectively, where 0 indicates dark and 2 indicates bright, and level 1 is the middle gray level.
- The filtering in the MATLAB simulation is achieved by Fourier transforming the input pattern, multiplying by a filter function, and taking an

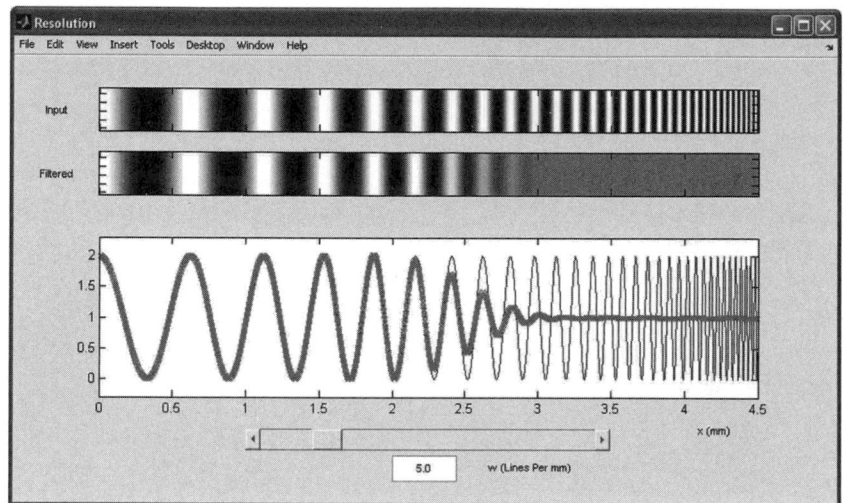

**FIGURE A.10**   Screen capture of the plot generated by running the MATLAB simulation "Resolution_GUI.m."

inverse Fourier transform. The choice of filter is a rectangular filter with smooth edges to generate a smooth filtering effect. This is only one of the many types of spatial filters that can affect image quality and reduce resolution.

• An example of image capture is shown in Figure A.10.

# Appendix B: Computing Diffraction Pattern

Light diffraction can be computed using the Fresnel (near field) and Fraunhofer (far field) formulations. The diffraction pattern can be calculated (in x-direction) from

$$U(x_0) = \iint h(x_0, x_1) \, U(x_1) dx_1 \tag{B.1}$$

where $U(x_0)$ is the aperture function (in x-direction), and

$$h(x_0, x_1) = \frac{\exp(jkz)}{j\lambda z} \exp\left[ j\frac{k}{2z}(x_0 - x_1)^2 \right] \tag{B.2}$$

The coordinate system for diffraction formulation is shown in Figure B.1. When the distance z is such that (in x-direction)

$$z^3 \gg \frac{\pi}{4\lambda}\left[(x_0 - x_1)^2\right]^2_{\text{Max}} \tag{B.3}$$

Equation (B.1) and Equation (B.2) can be approximated by (Fresnel approximation)

$$U(x_o) = \frac{\exp(jkz)}{j\lambda z} \exp\left[ j\frac{k}{2z}(x_o^2) \right] \int\int_{-\infty}^{+\infty} U(x_1) \exp\left[ j\frac{k}{2z}(x_1^2) \right] \exp\left[ -j\frac{2\pi}{\lambda z}(x_o x_1) \right] dx_1 \tag{B.4}$$

Equation (B.4) can be further simplified when the distance is such that (Fraunhofer condition) [1]

$$z \gg \frac{k\left(x_1^2\right)_{\text{Max}}}{2} \tag{B.5}$$

Then Equation (B.4) becomes

$$U(x_o) = \frac{\exp(jkz)}{j\lambda z} \exp\left[ j\frac{k}{2z}(x_o^2) \right] \int\int_{-\infty}^{+\infty} U(x_1) \exp\left[ -j\frac{2\pi}{\lambda z}(x_o x_1) \right] dx_1 \tag{B.6}$$

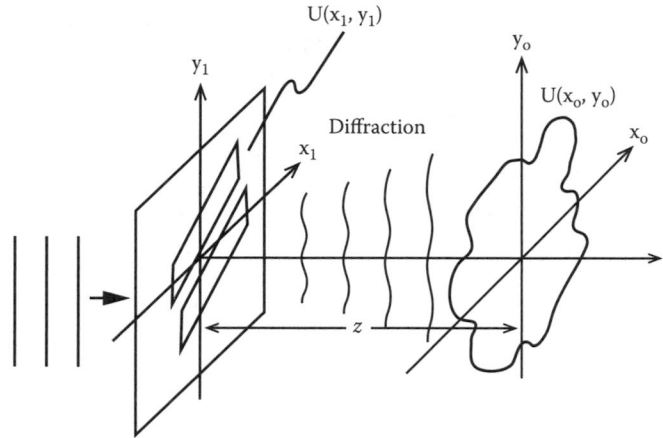

**FIGURE B.1**   Coordinate system for computing diffraction pattern.

Equations (B.4) to (B.6) are in one-dimension (x-coordinate) only. For two-dimensional (x-y) calculations, similar formulation follows [1], and the condition for Fresnel approximation becomes

$$z^3 \gg \frac{\pi}{4\lambda}\left[(x_0 - x_1)^2 + (y_0 - y_1)^2\right]^2_{\text{Max}} \tag{B.7}$$

and the condition for the Fraunhofer approximation becomes

$$z \gg \frac{k\left(x_1^2 - y_1^2\right)_{\text{Max}}}{2} \tag{B.8}$$

Equation (B.6) has the form of a Fourier transform (FT). Namely, $U(x_o)$ is the FT of $U(x_1)$, multiplied by the factors outside of the integral. Therefore far field (Fraunhofer) diffraction can be approximated by taking the Fourier transform of the object pattern, at spatial frequency $f_x = x_o/\lambda z, f_y = y_o/\lambda z$. Fourier transform in the digital domain, such as using MATLAB®, has the form

$$X(\kappa) = \sum_{n=1}^{N} x(n) e^{-j2\pi(k-1)\left(\frac{h-1}{n}\right)} \qquad 1 \le k \le n \tag{B.9}$$

where $X(\kappa)$ is the discrete Fourier transform of $x(n)$. In MATLAB the fast Fourier transform function (*fft*) is a valuable tool to compute diffraction patterns. The slit diffraction simulations utilize *fft* as a means of computing diffraction.

Near-field (Fresnel) diffraction (Equation B.4) can also be calculated using a Fourier transform. The difference between the Fresnel and Fraunhofer formulations is that the additional oscillating phase function multiplies the aperture function. This oscillating phase function converges to unity for far-field (Fraunhofer) diffraction, and therefore that term is ignored. To compute Fresnel diffraction using the *fft* routine, simply multiply the aperture function by the quadratic phase function

$$\exp\left[ j\frac{k}{2z}\left(x_1^2\right) \right]$$

and take the Fourier transform. Also note that the exponential functions outside the integrals in Equation (B.4) and Equation (B.6) are converted to unity when calculating the intensity by multiplying $U(x_0)$ by its complex conjugate.

## REFERENCE

1. Goodman, J. W. 2004. *Introduction to Fourier Optics*. 3rd ed. Greenwood Village, CO: Roberts & Company.

# Appendix C: Polarization Calculations Using Jones Vectors and Matrices

One of the methods for calculating polarization state is Jones vectors and matrices [1–3]. The polarization state is given by a vector, $E$, and the polarization component, such as a polarizer or a wave plate, is represented by a matrix, $A$. The transmitted light polarization state $E_t$ is given by

$$E_t = AE_i \tag{C.1}$$

where $E_i$ is the incident light polarization state. In fact, when there are multiple components, then the final output will be

$$E_t = A_N \cdots A_3 A_2 A_1 E_i \tag{C.2}$$

The polarization state is given by the Jones vector

$$E = \begin{bmatrix} E_{ox} \exp(i\phi_x) \\ E_{oy} \exp(i\phi_y) \end{bmatrix} \tag{C.3}$$

where $E_{ox}$ and $E_{oy}$ are the x and y electric field components, components of the polarization state, and $\phi_x$ and $\phi_y$ are the phases. Examples of Jones vectors representing linear and circular polarization are shown in Table C.1.

For a general linear polarizer, the matrix is given by

$$A_{pol} = \begin{bmatrix} \cos^2(\theta) & \cos(\theta)\sin(\theta) \\ \cos(\theta)\sin(\theta) & \sin^2(\theta) \end{bmatrix} \tag{C.4}$$

where $\theta$ is the polarizer angle. For example, a polarizer oriented horizontal ($\theta = 0$), vertical ($\theta = 90°$), and at $\theta = +45°$ is represented by the following:

$$A_{Pol\_Horiz} = \begin{bmatrix} 1 & 0 \\ 0 & 0 \end{bmatrix} \tag{C.5}$$

$$A_{Pol\_Vert} = \begin{bmatrix} 0 & 0 \\ 0 & 1 \end{bmatrix} \tag{C.6}$$

$$A_{Pol\_+45} = \begin{bmatrix} 1/2 & 1/2 \\ 1/2 & 1/2 \end{bmatrix} \tag{C.7}$$

---

**TABLE C.1**

**Jones Vectors for Selected Polarization States**

| Polarization State | Jones Vector |
|---|---|
| Horizontal | $\begin{bmatrix} 1 \\ 0 \end{bmatrix}$ |
| Vertical | $\begin{bmatrix} 0 \\ 1 \end{bmatrix}$ |
| +45° | $\dfrac{1}{\sqrt{2}} \begin{bmatrix} 1 \\ 1 \end{bmatrix}$ |
| Right circular | $\dfrac{1}{\sqrt{2}} \begin{bmatrix} 1 \\ -i \end{bmatrix}$ |
| Left circular | $\dfrac{1}{\sqrt{2}} \begin{bmatrix} 1 \\ i \end{bmatrix}$ |

*Note:*  Horizontal direction refers to x-direction, and vertical direction refers to y-direction.

---

For a generalized linear retarder, the Jones matrix is given by [2]

$$A_{ret} = \begin{bmatrix} e^{i\delta/2}\cos^2(\rho) + e^{-i\delta/2}\sin^2(\rho) & 2i\sin(\rho)\cos(\rho)\sin(\delta/2) \\ 2i\sin(\rho)\cos(\rho)\sin(\delta/2) & e^{-i\delta/2}\cos^2(\rho) + e^{i\delta/2}\sin^2(\rho) \end{bmatrix} \tag{C.8}$$

where $\delta$ is the phase retardation, and $\rho$ is the retarder axis orientation. For a quarter-wave plate $\delta = \pi/2$. For a half-wave plate $\delta = \pi$.

For example, for a quarter-wave plate ($\delta = \pi/2$) oriented at $\rho = -45°$, the transformation matrix becomes

$$A_{Quarter\_Wave\_+45} = \frac{1}{\sqrt{2}} \begin{bmatrix} 1 & -i \\ -i & 1 \end{bmatrix} \tag{C.9}$$

For example, for a half-wave plate ($\delta = \pi$) oriented at $\rho = -45°$, the transformation matrix becomes [3]

$$A_{Half\_Wave\_-45} = \begin{bmatrix} 0 & -i \\ -i & 0 \end{bmatrix} \tag{C.10}$$

Therefore the incident vertically polarized light passing through a half-wave plate rotates to horizontal polarization.

Other polarization components are also represented by Jones matrices, and listed by Hecht [1] and Kliger et al. [2].

## REFERENCES

1. Hecht, E. 2001. *Optics.* 4th ed. Reading, MA: Addison-Wesley.
2. Kliger, D. S., J. W. Lewis, and C. E. Randall. 1990. *Polarized Light in Optics and Spectroscopy.* Boston: Harcourt Brace Jovanovich.
3. Yariv, A., and P. Yeh. 2002. *Optical Waves in Crystals: Propagation and Control of Laser Radiation.* New York: Wiley Interscience.

# Appendix D: MATLAB®
# Simulation Function Files

This appendix contains the contents of the function or script file used in MATALB® simulation.

A word of caution: These simulations are to help the reader visualize and understand optical concept. They are not guaranteed to be error free, and they are not intended to be used as a ray tracing or optical design software. These simulations are not a replacement for commercially available software. Additionally, these simulations are not a substitute for professional help for designing optical components and systems.

## SIMULATION D.1: BLACKBODY RADIATION

Simulation file: Blackbody_radiation.m
Purpose: To become familiar with emission spectra of hot objects
Related chapter: 2–Light Sources

### FILE CONTENT

```
% Blackbody_radiation (Normalized blackbody radiation emission)
% Copyright Araz Yacoubian 2013, 2014

clear

close('all')

T = input('Input temperature in deg. Kelvin: ');
% temperature in degree Kelvin

h = 6.62606885E-34;% Plank's Constant (J.s)
c = 3.00E+08; % Speed of light (m/s)
k = 1.38E-23; % Boltzman constant (J/K)

lambda_set1 = 100:10:2000;
lambda_set2 = 2100:100:1000000;
lambda_nm = [lambda_set1 lambda_set2];%Wavelength (nm)
lambda_m = lambda_nm*(1e-9);     % Wavelength (meter)
lambda_um = lambda_nm*(1e-3);     % Wavelength (micrometer)

Param1_m = ((2*h*(c^2))./(lambda_m.^5));
Param2_m = 1./((exp((h*c/(k*T))./lambda_m))-1);
I = Param1_m.*Param2_m;
I_relative = I./(max(I));
```

```
[tmp,size_lambda] = size(lambda_nm);
[Y,lamda_max_index] = max(I_relative);
for m = lamda_max_index:size_lambda;
    if I_relative(m) < 0.05;
        max_index = m;
        break
    end
end

figure('Name','Blackbody Radiation','NumberTitle','off')
clf

axes('Fontsize',14)
plot(lambda_um,I_relative,'k-','LineWidth',2)
ylabel('Normalized Exitance','Fontsize',16)
xlabel('Wavelength (microns)','Fontsize',16)
axis([0 lambda_um(max_index) 0 1.1]);
axis([0 lambda_um(max_index) 0 1.1]);
grid
```

## SIMULATION D.2: REFRACTION

Simulation file: Refraction.m
Purpose: To become familiar with principle of refraction
Related chapter: 4–Manipulation of Light

### FILE CONTENT

```
% Refraction
% Copyright Araz Yacoubian 2013, 2014

clear

close('all')

n1 = input('Enter Refractive index in medium 1 (n1): ');
n2 = input('Enter Refractive index in medium 2 (n2): ');
q1_d = input('Input angle (deg.): ');
q1_r = (pi/180)*q1_d;
q2_r_out = asin((n1/n2)*sin(q1_r));

%%%%
if abs(imag(q2_r_out)) > 0;
    q2_r = -q1_r;
    x2 = -10;
else
    q2_r = q2_r_out;
    x2 = 10;
end

q2_d = (180/pi)*q2_r;
```

```
disp(' ');
disp(['Output angle = ' num2str(q2_d) ' (deg.)']);
%%%%


x1 = -10;
y1 = -x1*tan(q1_r);

x0 = 0;
y0 = 0;

y2 = -x2*tan(q2_r);

xin = [x1 x0];
yin = [y1 y0];
xout = [x0 x2];
yout = [y0 y2];

figure('Name','Refraction','NumberTitle','off')
plot(xin,yin,xout,yout,'— ')
line([0,0],[-10 10],'Color',[0 0 0])
line([-10 10],[0,0],'LineStyle',':','Color',[0 0 0])
axis([-10 10 -10 10])
xlabel('x (mm)')
ylabel('y (mm)')

text(-6,-8,['n1 = ' num2str(n1)],'BackgroundColor',[1 1 1],
'fontsize',14)
text(6,-8,['n2 = ' num2str(n2)],'BackgroundColor',[1 1 1],
'fontsize',14)
```

## SIMULATION D.3: REFRACTION (GUI)

Simulation file: Refraction_GUI.m
Purpose: To become familiar with principle of refraction; same as Refraction.m,
however uses graphic user interface (GUI)
Related chapter: 4–Manipulation of Light

### FILE CONTENT

```
function Refraction_GUI
% Refraction GUI
% Copyright Araz Yacoubian 2013, 2014
clear
close('all')

fh = figure('Visible','off');

set(fh,'Color',[0.925 0.914 0.847]);

set(fh,'Visible','on','Name','Refraction','NumberTitle','off');
```

```
%%%%%%%%%%%%%
slider_value = 20;
angle_input = slider_value;
n1 = 1;
n2 = 1.5;
%%%%%%%%%%%%%

CalcPlot1

angle_input_string = num2str(angle_input,'%2.1f');
text_angle_input = uicontrol('Style','edit',...
        'String',angle_input_string,'Position',[60,210,35,25],...
        'BackgroundColor',[1 1 1]);%,...

%%%%%%%%%%%%%%%%%%%%%%%%%%%%%
sh = uicontrol(fh,'Style','slider',...
          'Max',90,'Min',-90,'Value',slider_value,...
          'SliderStep',[0.05 0.2],...
          'Position',[30 100 20 250],...
          'Callback',{@slider1_Callback});

function slider1_Callback(source,eventdata)
   slider_value = get(source, 'Value');
   slider_value_dsp = num2str(slider_value,'%2.1f');
   set(textinputSliderValue, 'String', slider_value_dsp);

   CalcPlot1

end

angle_input_string = num2str(angle_input,'%2.1f');
textinputSliderValue = uicontrol('Style','edit',...
        'Value',angle_input,'Position',[60,210,35,25],...%[140,
        round(5.7*btn_pd_y),100,25],...
        'BackgroundColor',[1 1 1],...
        'String',angle_input_string,...
        'Callback',{@anglein_Callback});

function anglein_Callback(source,eventdata)
  angle_input = str2double(get(source, 'String'));
  angle_input
  slider_value = angle_input;

sh = uicontrol(fh,'Style','slider',...
        'Max',90,'Min',-90,'Value',slider_value,...
        'SliderStep',[0.05 0.2],...
        'Position',[30 100 20 250],...
        'Callback',{@slider1_Callback});

   CalcPlot1
```

```
end
%%%%%%%%%%%%%%%%%%%%%%%%%

%%%%%%%%%%%%%%%%%%%%%%%%%%%%%%%%%%%%
text_n1_mark = uicontrol('Style','text',...
        'String','n1 = ','Position',[180,20,85,25],...
        'BackgroundColor',[0.925 0.914 0.847]);

function textin_n1(source,eventdata)
  n1 = str2double(get(source, 'String'));
  CalcPlot1
end

n1_string = num2str(n1,'%2.1f');
text_n1 = uicontrol('Style','edit',...
        'String',n1_string,'Position',[240,25,25,25],...
        %[160,25,25,25],...
        'BackgroundColor',[1 1 1],...
        'Value',n1,'Callback',{@textin_n1});
%%%%%%%%%%%%%%%%%%%%%%%%%%%%%%%%%%%%%

%%%%%%%%%%%%%%%%%%%%%%%%%%%%%%%%%%%%
text_n2_mark = uicontrol('Style','text',...
        'String','n2 = ','Position',[330,20,85,25],...
        'BackgroundColor',[0.925 0.914 0.847]);

function textin_n2(source,eventdata)
  n2 = str2double(get(source, 'String'));
  CalcPlot1
end

n2_string = num2str(n2,'%2.1f');
text_n2 = uicontrol('Style','edit',...
         'String',n2_string,'Position',[390,25,25,25],...
         %[360,25,25,25],...
        'BackgroundColor',[1 1 1],...
        'Value',n2,'Callback',{@textin_n2});

%%%%%%%%%%%%%%%%%%%%%%%%%%%%%%%%%%%%%%%%
%%%%%%%%%%%

text_angle_mark = uicontrol('Style','text',...
        'String','q1',...
        'FontName','Symbol','FontSize',11,...
        'Position',[65,235,25,25],...
        'BackgroundColor',[0.925 0.914 0.847]);
%%%%%%%%%%%

%%%%%%%%%%%%%%%
function CalcPlot1
q1_d = slider_value;
```

```
q1_r = (pi/180)*q1_d;
q2_r_out = asin((n1/n2)*sin(q1_r));

if abs(imag(q2_r_out)) > 0;
   q2_r = -q1_r;
   x2 = -10;
else
   q2_r = q2_r_out;
   x2 = 10;
end

q2_d = (180/pi)*q2_r;

disp(['Input angle = ' num2str(q1_d) ' (deg.)']);
disp(['Output angle = ' num2str(q2_d) ' (deg.)']);
disp(' ');
%%%%
x05 = -5;
x1 = -10;
y05 = -x05*tan(q1_r);
y1 = -x1*tan(q1_r);

x0 = 0;
y0 = 0;
y2 = -x2*tan(q2_r);
x_lft = [x1 x05];
y_lft = [y1 y05];
x = [x1 x0 x2];
y = [y1 y0 y2];

xin = [x1 x0];
yin = [y1 y0];
xout = [x0 x2];
yout = [y0 y2];

xpos = 120;
ypos = 90;
pltx_wd = 390;
plty_wd = 300;

a_ax = axes('Units','pixels','Position',[xpos,ypos,pltx_
wd,plty_wd], 'XTickLabel','','YTickLabel','');

plot(xin,yin,xout,yout,'- ')
axis('equal')
line([0,0],[-10 10],'Color',[0 0 0])
line([-10 10],[0,0],'LineStyle',':','Color',[0 0 0])
axis([-10 10 -10 10])
xlabel('x (mm)')
ylabel('y (mm)')
```

```
text(-6,-8,['n1 = ' num2str(n1)],'BackgroundColor',[1 1 1],
'fontsize',12)
text(6,-8,['n2 = ' num2str(n2)],'BackgroundColor',[1 1 1],
'fontsize',12)

end

end
```

## SIMULATION D.4: DIFFRACTION (GUI)

Simulation file: Diffraction_GUI.m
Purpose: To become familiar with principle of diffraction
Related chapter: 4–Manipulation of Light

### FILE CONTENT

```
function Diffraction_GUI
% Diffraction
% Copyright Araz Yacoubian 2013, 2014

clear

close('all')

fh = figure('Visible','off');
set(fh,'Units','Normalized','Position',[0.05 0.43 0.44 0.42]);

set(fh,'Color',[0.925 0.914 0.847]);
% set(fh,'Visible','on');
set(fh,'Visible','on','Name','Diffraction','NumberTitle','off');

fwndw_H = 565;
fwndw_V = 420;

xpos2_n = 120/fwndw_H;
ypos2_n = 270/fwndw_V;
pltx_wd2_n = 390/fwndw_H;
plty_wd2_n = 120/fwndw_V;

a_ax1 = axes('Units','Normalized','Position',...
  [xpos2_n,ypos2_n,pltx_wd2_n,plty_wd2_n],'XTickLabel','','YTi
  ckLabel','');
set(a_ax1,'Parent',fh);
%%%%%%%%%%%%%%%%%%

xpos2B = 120;
ypos2B = 270-150;
```

```
pltx_wd2B = 390;
plty_wd2B = 120;

xpos2B_n = xpos2B/fwndw_H;
ypos2B_n = ypos2B/fwndw_V;
pltx_wd2B_n = pltx_wd2B/fwndw_H;
plty_wd2B_n = plty_wd2B/fwndw_V;

a_ax2B = axes('Units','Normalized','Position',...
  [xpos2B_n,ypos2B_n,pltx_wd2B_n,plty_wd2B_n],'XTickLabel','
  ','YTickLabel','');
set(a_ax2B,'Parent',fh);

%%%%%%%%%%%%%%
slider_valueDist = 1000;
slider_valueDiste = log(slider_valueDist);
z_mm = slider_valueDist;

z_mm_string = num2str(z_mm,'%2.1f');
text_z_mm = uicontrol('Style','edit',...
        'String',z_mm_string,'Units','Normalized',...
        'Position',[20/fwndw_H,360/fwndw_V,45/fwndw_H,25/
        fwndw_V],...
        'BackgroundColor',[1 1 1]);
%%%%%%%%%%%%%%

%%%%%%%%%%%%%%
wh = 0.2;
slider_value_wh = 0.2;

wh_string = num2str(wh,'%2.1f');
text_wh = uicontrol('Style','edit',...
        'String',wh_string,'Units','Normalized',...
        'Position',[160/fwndw_H,40/fwndw_V,45/fwndw_H,25/
        fwndw_V],...
        'BackgroundColor',[1 1 1]);

text_wh_mark = uicontrol('Style','text',...
        'String','Slit Half Width (mm)','Units','Normalized',...
        'Position',[110/fwndw_H,30/fwndw_V,45/fwndw_H,35/
        fwndw_V],...
        'BackgroundColor',[0.925 0.914 0.847]);

sh_wh = uicontrol(fh,'Style','slider',...
        'Max',5,'Min',0.1,'Value',slider_value_wh,...
        'SliderStep',[0.1 0.2],...
        'Units','Normalized',...
        'Position',[100/fwndw_H 70/fwndw_V 250/fwndw_H 20/
        fwndw_V],...
        'Callback',{@slider_wh_Callback});
```

```
function slider_wh_Callback(source,eventdata)
   slider_value_wh = get(source, 'Value');
   slider_value_wh_dsp = num2str(slider_value_wh,'%2.1f');
   set(textinputSliderValue2wh, 'String', slider_value_wh_dsp);
   wh = slider_value_wh;
   Ufrl_calc
end


wh_input_string = num2str(wh,'%2.1f');
textinputSliderValue2wh = uicontrol('Style','edit',...
        'Value',wh,...
        'Units','Normalized',...
        'Position',[160/fwndw_H,40/fwndw_V,45/fwndw_H,25/
        fwndw_V],...
        'BackgroundColor',[1 1 1],...
        'String',wh_input_string,...
        'Callback',{@wh_Callback});
function wh_Callback(source,eventdata)
  wh = str2double(get(source, 'String'));

  slider_value_wh = wh;

sh_wh = uicontrol(fh,'Style','slider',...
        'Max',5,'Min',0.1,'Value',slider_value_wh,...
        'SliderStep',[0.1 0.2],...
        'Units','Normalized',...
        'Position',[100/fwndw_H 70/fwndw_V 250/fwndw_H 20/
        fwndw_V],...
        'Callback',{@slider_wh_Callback});

        Ufrl_calc
end
%%%%%%%%%%%%%%

%%%%%%%%%%%%%%
L = 1;
slider_value_L = 1;

L_string = num2str(L,'%2.1f');
text_L = uicontrol('Style','edit',...
     'String',L_string,...
     'Units','Normalized',...
     'Position',[340/fwndw_H,10/fwndw_V,45/fwndw_H,25/fwndw_V],...
     'BackgroundColor',[1 1 1]);%,...

sh_L = uicontrol(fh,'Style','slider',...
        'Max',5,'Min',0.5,'Value',slider_value_L,...
        'SliderStep',[0.1 0.2],...
        'Units','Normalized',...
```

```matlab
        'Position',[240/fwndw_H 40/fwndw_V 250/fwndw_H 20/
        fwndw_V],...
        'Callback',{@slider_L_Callback});

function slider_L_Callback(source,eventdata)
  slider_value_L = get(source, 'Value');
  slider_value_L_dsp = num2str(slider_value_L,'%2.1f');
  set(textinputSliderValue2L, 'String', slider_value_L_dsp);
  L = slider_value_L;
  Ufrl_calc

end

text_L_mark = uicontrol('Style','text',...
        'String','Slit Separation (Center-Center, mm) ',...
        'Units','Normalized',...
        'Position',[380/fwndw_H,5/fwndw_V,180/fwndw_H,25/
        fwndw_V],...
        'BackgroundColor',[0.925 0.914 0.847]);

text_L2_mark = uicontrol('Style','text',...
        'String','x (mm)',...
        'Units','Normalized',...
        'Position',[380/fwndw_H,70/fwndw_V,200/fwndw_H,25/
        fwndw_V],...
        'BackgroundColor',[0.925 0.914 0.847]);

L_input_string = num2str(L,'%2.1f');
textinputSliderValue2L = uicontrol('Style','edit',...
        'Value',L,...
        'Units','Normalized',...
        'Position',[340/fwndw_H,10/fwndw_V,45/fwndw_H,25/
        fwndw_V],...
        'BackgroundColor',[1 1 1],...
        'String',L_input_string,...
        'Callback',{@L_Callback});

function L_Callback(source,eventdata)

  L = str2double(get(source, 'String'));

  slider_value_L = L;

sh_L = uicontrol(fh,'Style','slider',...
        'Max',5,'Min',0.1,'Value',slider_value_L,...
        'SliderStep',[0.1 0.2],...
        'Units','Normalized',...
        'Position',[100/fwndw_H 70/fwndw_V 250/fwndw_H 20/
        fwndw_V],...
        'Callback',{@slider_L_Callback});

        Ufrl_calc
```

```
end
%%%%%%%%%%%%%%

lambda_micron = 0.633;     % Wavelength (microns)
lambda = lambda_micron*10^(-3);% Wavelength (mm)

xmax = 50;
xstp = 0.001;
x = -xmax:xstp:xmax-xstp;
[tmp,szx] = size(x);

xf = (1/(1*xstp)*(1/szx)).*((-szx/2):(szx/2)-1);

x2 = (lambda*z_mm).*xf;

midx = round(szx/2);

xh = round(wh/xstp);

xL = round(L/xstp);

k_1d = 2*pi/lambda;

%%%%%%%%%%%%%%%%%%%%%%%%%%%

sh = uicontrol(fh,'Style','slider',...
        'Max',13.5,'Min',1.15,'Value',slider_valueDiste,...
        'SliderStep',[0.05 0.2],...
        'Units','Normalized',...
        'Position',[30/fwndw_H 100/fwndw_V 20/fwndw_H 250/
        fwndw_V],...
        'Callback',{@slider1_Callback});

%%%%%%%%%%%%%%%%%%%%%%%%%%%%%%%%%%%%%%%

text_lambda_mark = uicontrol('Style','text',...
        'String','Wavelength',...
        'Units','Normalized',...
        'Position',[7/fwndw_H,50/fwndw_V,90/fwndw_H,25/
        fwndw_V],...
        'BackgroundColor',[0.925 0.914 0.847]);

text_lambda_mark = uicontrol('Style','text',...
        'String','(microns) ',...
        'Units','Normalized',...
        'Position',[8/fwndw_H,35/fwndw_V,90/fwndw_H,25/
        fwndw_V],...
        'BackgroundColor',[0.925 0.914 0.847]);

function textin_lambda(source,eventdata)
```

```
  lambda_micron = str2double(get(source, 'String'));
  lambda = lambda_micron*10^(-3);% Wavelength (mm)
  Ufrl_calc
end

lambda_micron_string = num2str(lambda_micron,'%2.3f');

text_lambda = uicontrol('Style','edit',...
      'String',lambda_micron_string,...
      'Units','Normalized',...
      'Position',[20/fwndw_H,15/fwndw_V,55/fwndw_H,25/
      fwndw_V],...
      'BackgroundColor',[1 1 1],...
      'Value',lambda_micron,'Callback',{@textin_lambda});
      lambda = lambda_micron*10^(-3);% Wavelength (mm)

%%%%%%%%%%%%%%%%%%%%%%%%%%%%%%%%%%%%%%%%

function slider1_Callback(source,eventdata)
  slider_valueDiste = get(source, 'Value');
  slider_valueDist = exp(slider_valueDiste);
  slider_value_dsp1 = num2str(slider_valueDist,'%2.1f');
  set(textinputSlidervalueDist, 'String', slider_value_dsp1);
  Ufrl_calc
  slider_valueDist = exp(slider_valueDiste);
end

z_mm_input_string = num2str(z_mm,'%2.1f');
textinputSlidervalueDist = uicontrol('Style','edit',...
      'Value',z_mm,...
      'Units','Normalized',...
      'Position',[20/fwndw_H,360/fwndw_V,45/fwndw_H,25/
      fwndw_V],...
      'BackgroundColor',[1 1 1],...
      'String',z_mm_input_string,...
      'Callback',{@z_mm_Callback});

text_z_mm_mark = uicontrol('Style','text',...
      'String','Distance (mm) ',...
      'Units','Normalized',...
      'Position',[10/fwndw_H,385/fwndw_V,80/fwndw_H,25/
      fwndw_V],...
      'BackgroundColor',[0.925 0.914 0.847]);
function z_mm_Callback(source,eventdata)
  z_mm = str2double(get(source, 'String'));

  slider_valueDist = z_mm;

  sh = uicontrol(fh,'Style','slider',...
      'Max',13.5,'Min',1.15,'Value',slider_valueDiste,...
      'SliderStep',[0.05 0.2],...
```

```
        'Units','Normalized',...
        'Position',[30/fwndw_H 100/fwndw_V 20/fwndw_H 250/
        fwndw_V],...
        'Callback',{@slider1_Callback});
        z_mm = exp(slider_valueDiste);

        Ufrl_calc
end
%%%%%%%%%%%%%%%%%%%%%%%%%%%%%%

zne = 0;
rng_val_Fra = (2*z_mm)/(k_1d*(wh^2));

tmpwmax = [L+2*wh wh];
wMax = max(tmpwmax);
rng_val_Frl = (4*lambda*(z_mm^3))/((wMax)^4);

if rng_val_Frl > 35;
    zne = 1; % Fresnel
end

if rng_val_Fra > 30;
    zne = 2; % Fraunhofer
end

if zne < 1
    disp('Not in Fresnel or Fraunhofer Zone')
    disp('Try larger z')
    disp(' ')
return
end
%%% End. Zone Calc%%%

Uo = 0.*x;
Uo(midx - round(xL/2) - xh : midx - round(xL/2) + xh) = 1;
Uo(midx + round(xL/2) - xh : midx + round(xL/2) + xh) = 1;
% end

%%%%%%%%%%% Fresnel
  %%%%%%%%%% Beg. Ufrl_calc%%%%%%%%

  %%%%%%%%%%%%%%%%%%
Ufrl_calc

function Ufrl_calc

z_mm = slider_valueDist;
k_1d = 2*pi/lambda;
xh = round(wh/xstp);
xL = round(L/xstp);
```

```
%%%%% Zone Calc%%%

zne = 0;

rng_val_Fra = (2*z_mm)/(k_1d*(wh^2));

rng_val_Frl = (4*lambda*(z_mm^3))/((wh)^4);

if rng_val_Frl > 35;
   zne = 1; % Fresnel
end
if rng_val_Fra > 30;
   zne = 2; % Fraunhofer
end

disp(' ')

   if zne < 1
   disp('Not in Fresnel or Fraunhofer Zone')
   disp('Try larger z')
   disp(' ')
return
end

%%% End. Zone Calc%%%

Uo = 0.*x;
Uo(midx - round(xL/2) - xh : midx - round(xL/2) + xh) = 1;
Uo(midx + round(xL/2) - xh : midx + round(xL/2) + xh) = 1;

U_preFFT = Uo.*exp((i*k_1d/(2*z_mm)).*(x.^2));

Ufrl = fft(U_preFFT);
Ufrl = fftshift(Ufrl);

Ufrl_max = max(abs(Ufrl));
Ufrlmx1 = (1/Ufrl_max).*Ufrl;

x2 = (lambda*z_mm).*xf;

x2limit = 1000*(lambda*(sqrt(z_mm)))+0.8;

I_diff = (abs(Ufrl)).^2;
I_diff_max = max(I_diff);
I_diff_norm = (1/I_diff_max).*I_diff;

%%%%%%%%%%%%%%%%%%%%%%%%%%%%%%%%%%%%%

axes(a_ax1)
```

```
plot(x2,abs(Ufrl))
axis([-x2limit x2limit -0.1*max(abs(Ufrl)) 1.1*max(abs(Ufrl))]);
plot(x2,I_diff_norm)
axis([-x2limit x2limit -0.1 1.1]);
ylabel('Intensity')


%%%%%%%%%%%%%%%%%%%%%%%%%%%%%%%%%%%%%%%%
%%%%%%%%%%%%%%%%%%%%%%%%%%%%%%%%%%%%%%%%


axes(a_ax2B);
plot(x,Uo)
axis([-1.5*(2*wh+L) 1.5*(2*wh+L) -0.1 1.1]);


%%%%%%%%%%%%%%%%%%%%%%%%%%%%%%%%%%%%%%%%%


end

end
```

## SIMULATION D.5: POLARIZATION

Purpose: To familiarize with polarizers, quarter- and half-wave plates
Related chapter: 5–Polarization
Simulation file: Polarization_GUI.m

### FILE CONTENT

```
function Polarization_GUI
% Polarization
% Copyright Araz Yacoubian 2013, 2014

clear
close('all')

%%%%
option_wv_pol = 21;% In, qrtwv

%%%%

x = -125:125;

q1 = 30;45;

q2 = 135;

r1 = 50;
r2 = 55;

sgm = pi/2;
```

```
sgm = pi/1;
rho = 50;

%%%%%%%%%%%%%

fh303 = figure('Visible','off');
set(fh303,'Units','Normalized','Position',[0.2 0.45 0.6 0.45]);

set(fh303,'Color',[0.925 0.914 0.847]);
set(fh303,'Visible','on','Name','Polarization','NumberTitle',
'off');

%%%%%%%%%%%%%

%%%%%%%%%%%%%%
slider_value_q1 = q1;

q1_string = num2str(q1,'%2.1f');
text_q1 = uicontrol('Style','edit',...
        'String',q1_string,...
        'Units','Normalized',...
        'Position',[0.09,0.25,0.05,0.05],...
        'BackgroundColor',[1 1 1]);

sh_q1 = uicontrol(fh303,'Style','slider',...
        'Max',360,'Min',0,'Value',slider_value_q1,...
        'SliderStep',[0.0500 0.20],...
        'Units','Normalized',...
        'Position',[0.17,0.25,0.3,0.04],...
        'Callback',{@slider_q1_Callback});

function slider_q1_Callback(source,eventdata)
  slider_value_q1 = get(source, 'Value');
  slider_value_q1_dsp = num2str(slider_value_q1,'%2.1f');
  set(textinputSliderValue2q1, 'String', slider_value_q1_dsp);
  q1 = slider_value_q1;
  pol_calc

end

text_q1_mark = uicontrol('Style','text',...
        'String','P1 Angle',...
        'Units','Normalized',...
        'Position',[0.005,0.25,0.06,0.04],...
        'BackgroundColor',[0.925 0.914 0.847]);

q1_input_string = num2str(q1,'%2.1f');
textinputSliderValue2q1 = uicontrol('Style','edit',...
        'Value',q1,...
        'Units','Normalized',...
```

```
        'Position',[0.09,0.25,0.05,0.05],...
        'BackgroundColor',[1 1 1],...
        'String',q1_input_string,...
        'Callback',{@q1_Callback});

function q1_Callback(source,eventdata)
  q1 = str2double(get(source, 'String'));

  slider_value_q1 = q1;

      pol_calc
end
%%%%%%%%%%%%%

%%%%%%%%%%%%%

slider_value_rho = rho;

rho_string = num2str(rho,'%2.1f');
text_rho = uicontrol('Style','edit',...
        'String',rho_string,...
        'Units','Normalized',...
        'Position',[0.09,0.25-0.10,0.05,0.05],...
        'BackgroundColor',[1 1 1]);
sh_rho = uicontrol(fh303,'Style','slider',...
         'Max',360,'Min',0,'Value',slider_value_rho,...
         'SliderStep',[0.0500 0.20],...
         'Units','Normalized',...
         'Position',[0.17,0.25-0.10,0.3,0.04],...
         'Callback',{@slider_rho_Callback});

function slider_rho_Callback(source,eventdata)
  slider_value_rho = get(source, 'Value');
  slider_value_rho_dsp = num2str(slider_value_rho,'%2.1f');
  set(textinputSliderValue2rho, 'String', slider_value_rho_dsp);
  rho = slider_value_rho;
  pol_calc

end

text_rho_mark = uicontrol('Style','text',...
        'String','Wave Angle ',...
        'Units','Normalized',...
        'Position',[0.005,0.25-0.10,0.09,0.04],...
        'BackgroundColor',[0.925 0.914 0.847]);
%              'Position',[0.005,0.25-0.10*2,0.06,0.04],...
rho_input_string = num2str(rho,'%2.1f');
textinputSliderValue2rho = uicontrol('Style','edit',...
        'Value',rho,...
        'Units','Normalized',...
```

```
                  'Position',[0.09,0.25-0.10,0.05,0.05],...
                  'BackgroundColor',[1 1 1],...
                  'String',rho_input_string,...
                  'Callback',{@rho_Callback});

function rho_Callback(source,eventdata)
  rho = str2double(get(source, 'String'));

  slider_value_rho = rho;

      pol_calc

end
%%%%%%%%%%%%%%%
%%%%%%%%%%%%%%%

slider_value_q2 = q2;

q2_string = num2str(q2,'%2.1f');
text_q2 = uicontrol('Style','edit',...
        'String',q2_string,...
        'Units','Normalized',...
        'Position',[0.09,0.25-0.10*2,0.05,0.05],...
        'BackgroundColor',[1 1 1]);

sh_q2 = uicontrol(fh303,'Style','slider',...
           'Max',360,'Min',0,'Value',slider_value_q2,...
           'SliderStep',[0.0500 0.20],...
           'Units','Normalized',...
           'Position',[0.17,0.25-0.10*2,0.3,0.04],...
           'Callback',{@slider_q2_Callback});

function slider_q2_Callback(source,eventdata)
  slider_value_q2 = get(source, 'Value');
  slider_value_q2_dsp = num2str(slider_value_q2,'%2.1f');
  set(textinputSliderValue2q2, 'String', slider_value_q2_dsp);
  q2 = slider_value_q2;
  pol_calc

end

text_q2_mark = uicontrol('Style','text',...
        'String','P2 Angle',...
        'Units','Normalized',...
        'Position',[0.005,0.25-0.10*2,0.06,0.04],...
        'BackgroundColor',[0.925 0.914 0.847]);

q2_input_string = num2str(q2,'%2.1f');
textinputSliderValue2q2 = uicontrol('Style','edit',...
```

```
      'Value',q2,...
      'Units','Normalized',...
      'Position',[0.09,0.25-0.10*2,0.05,0.05],...
      'BackgroundColor',[1 1 1],...
      'String',q2_input_string,...
      'Callback',{@q2_Callback});

function q2_Callback(source,eventdata)
  q2 = str2double(get(source, 'String'));
  slider_value_q2 = q2;


      pol_calc
end
%%%%%%%%%%%%%%

pol_calc

%%%%%%%%%%%%%%%%
%%%%%%%%%%%%
  function pol_calc
yc1_p = real(sqrt(r1^2-x.^2));
yc1_n = -real(sqrt(r1^2-x.^2));

yc1_p_imag = imag(sqrt(r1^2-x.^2));
yc1_p_tmp = yc1_p;
yc1_p_tmp(find(yc1_p_imag>0)) = nan;
yc1_p = yc1_p_tmp;
yc1_n_tmp = yc1_n;
yc1_n_tmp(find(yc1_p_imag>0)) = nan;
yc1_n = yc1_n_tmp;

x1 = r1.*cos(q1*pi/180);
y1 = r1.*sin(q1*pi/180);
plot_x1 = [-x1 x1];
plot_y1 = [-y1 y1];
%%%%%%%%%%%%%%%%

%%%%%%%%%%%%%%%%%%%%%%%%
sftx_2 = 0;
sfty_2 = 0;

yc2_p = real(sqrt(r2^2-(x-sftx_2).^2))+sfty_2;
yc2_n = -real(sqrt(r2^2-(x-sftx_2).^2))+sfty_2;

yc2_p_imag = imag(sqrt(r2^2-(x-sftx_2).^2))+sfty_2;
yc2_p_tmp = yc2_p;
yc2_p_tmp(find(yc2_p_imag>sfty_2)) = nan;
yc2_p = yc2_p_tmp;
yc2_n_tmp = yc2_n;
```

```
yc2_n_tmp(find(yc2_p_imag>sfty_2)) = nan;
yc2_n = yc2_n_tmp;


x2 = r2.*cos(q2*pi/180);
y2 = r2.*sin(q2*pi/180);
plot_x2 = [-x2+sftx_2 x2+sftx_2];
plot_y2 = [-y2+sfty_2 y2+sfty_2];
%%%%%%%%%%%%%%%%%%%%%%%%%


%%%%%%%%%%%%%%%%%%%%
x_sq = 60;
y_sq = 35;
rho_rad_disp = -rho*pi/180;
A_rt = [cos(rho_rad_disp) sin(rho_rad_disp);...
  -sin(rho_rad_disp) cos(rho_rad_disp)];


V1 = [x_sq;y_sq];
Vout1 = A_rt*V1;


V2 = [-x_sq;y_sq];
Vout2 = A_rt*V2;
V3 = [-x_sq;-y_sq];
Vout3 = A_rt*V3;


V4 = [x_sq;-y_sq];
Vout4 = A_rt*V4;


Voutsq_x = [Vout1(1) Vout2(1) Vout3(1) Vout4(1) Vout1(1)];
Voutsq_y = [Vout1(2) Vout2(2) Vout3(2) Vout4(2) Vout1(2)];


r_rho = x_sq;
xW = r_rho.*cos(rho*pi/180);
yW = r_rho.*sin(rho*pi/180);
plot_xW = [-xW xW];
plot_yW = [-yW yW];
%%%%%%%%%%%%%%%%%%%%%%%%%


%%%%%%%%%%%%%%%%%%
Hoz_Ln_x = x;
Hoz_Ln_y = 0.*x;
Hoz_Ln_y(66:186) = NaN;


Ver_Ln_y = x;
Ver_Ln_x = 0.*x;
Ver_Ln_x(66:186) = NaN;


%%%%%%%%%%%%%%%%%%%%%%%%%
q1_rad = (pi/180)*q1;
E_in = [cos(q1_rad);sin(q1_rad)];
```

```
q2_rad = q2*(pi/180);
A_pol2 = [(cos(q2_rad))^2 sin(q2_rad)*cos(q2_rad);...
  sin(q2_rad)*cos(q2_rad) (sin(q2_rad))^2];

rho_rad = rho*pi/180;
A_rtrd = [(exp(i*sgm/2))*((cos(rho_rad))^2) + (exp(-i*sgm/2))*
((sin(rho_rad))^2)...
  2*i*(sin(rho_rad))*(cos(rho_rad))*sin(sgm/2);...
  2*i*(sin(rho_rad))*(cos(rho_rad))*sin(sgm/2)...
  (exp(-i*sgm/2))*((cos(rho_rad))^2) + (exp(i*sgm/2))*((sin(rho_
  rad))^2)];

%%%%%%%%%%%%%%%%%%%%%%%%%%%%

%%%%%%%%%%%%%%%%%%%%%%%%%%%%

tplot = -2*pi:(pi/50):2*pi;

%%%%%%%%%%%%%%%%%%%%%%%%%%%

%%%%%%%
bgh_WP = uibuttongroup('Parent',fh303,'Title','Component
Selection',...
  'Units','Normalized',...
  'Position',[.5.03.3 0.22]);

rbh_P2 = uicontrol(bgh_WP,'Style','radiobutton','String','P2 ',...
        'Units','normalized',...
        'Value',68,...
        'Position',[.7.5.3.5]);
rbh_Wq = uicontrol(bgh_WP,'Style','radiobutton','String'
,'1/4 ',...
        'FontName','Symbol',...
        'Units','normalized',...
        'Value',1,...
        'Position',[.4.5.3.5]);

rbh_Wh = uicontrol(bgh_WP,'Style','radiobutton','String','1/2
',...
        'FontName','Symbol',...
        'Units','normalized',...
        'Value',2,...
        'Position',[.1.5.3.5]);
rbh_Wq_P2 = uicontrol(bgh_WP,'Style','radiobutton','String',
'1/4,',...
        'FontName','Symbol',...
        'Units','normalized',...
        'Value',3,...
        'Position',[.5.03.2.5]);
```

```matlab
text_Wq_P2_b = uicontrol(bgh_WP,'Style','text',...
        'String','P2','Units','Normalized',...
        'Position',[.7.05.1.3]);

rbh_Wh_P2 = uicontrol(bgh_WP,'Style','radiobutton','String'
,'1/2,',...
        'FontName','Symbol',...
        'Units','normalized',...
        'Position',[.1.03.2.5]);

text_Wh_P2_b = uicontrol(bgh_WP,'Style','text',...
        'String','P2','Units','Normalized',...
        'Position',[.3.05.1.3]);

set(bgh_WP,'SelectionChangeFcn',@radiobutton_W_P_Callback);
set(bgh_WP,'Visible','on');
if option_wv_pol = = 1; % In, pol
    Slctn_deflt = [rbh_P2];
end

if option_wv_pol = = 21;% In, qrtwv
        Slctn_deflt = [rbh_Wq];
end

if option_wv_pol = = 22;% In, hlfwv
        Slctn_deflt = [rbh_Wh];
end

if option_wv_pol = = 31;% In, qrtwv, pol
        Slctn_deflt = [rbh_Wq_P2];
end
if option_wv_pol = = 32;% In, hlfwv, pol
        Slctn_deflt = [rbh_Wh_P2];
end

set(bgh_WP,'SelectedObject',Slctn_deflt);
%%%%%%%
function radiobutton_W_P_Callback(source,eventdata)

btn_slctd = get(get(source,'SelectedObject'),'String');

if btn_slctd = = '1/2 ';
    disp(['This is 1/2'])
    option_wv_pol = 22;
end

if btn_slctd = = '1/4 ';
    disp(['This is 1/4'])
    option_wv_pol = 21;
end

if btn_slctd = = 'P2 ';
    disp(['This P2'])
    option_wv_pol = 1;
```

```matlab
end

if btn_slctd = = '1/2,';
   disp(['This is 1/2, P2'])
   option_wv_pol = 32;
end

if btn_slctd = = '1/4,';
   disp(['This is 1/4, P2'])
   option_wv_pol = 31;
end

option_wv_pol
pol_calc

end

%%%%%%%%%
function A_rtrd_fnctn
A_rtrd = [(exp(i*sgm/2))*((cos(rho_rad))^2) + (exp(-i*sgm/2))*
((sin(rho_rad))^2)...
   2*i*(sin(rho_rad))*(cos(rho_rad))*sin(sgm/2);...
   2*i*(sin(rho_rad))*(cos(rho_rad))*sin(sgm/2)...
   (exp(-i*sgm/2))*((cos(rho_rad))^2) + (exp(i*sgm/2))*((sin(rho_
   rad))^2)];
end

if option_wv_pol = = 1;
  E_out_tot = A_pol2*E_in;
end

if option_wv_pol = = 21;
   sgm = pi/2;% Qrt_W = pi/2
   A_rtrd_fnctn;
   E_out_tot = A_rtrd*E_in;
end

if option_wv_pol = = 22;
   sgm = pi;% Hlf_W = pi
   A_rtrd_fnctn;
   E_out_tot = A_rtrd*E_in;
end

if option_wv_pol = = 31;
   sgm = pi/2;% Qrt_W = pi/2
   A_rtrd_fnctn;
   E_out_tot = A_pol2*A_rtrd*E_in;
end

if option_wv_pol = = 32;
   sgm = pi;% Hlf_W = pi
```

```
    A_rtrd_fnctn;
    E_out_tot = A_pol2*A_rtrd*E_in;
end

%%%%%%%%% V6

E_out_tot_x = E_out_tot(1);
E_out_tot_y = E_out_tot(2);

E_out_tot_x_abs = abs(E_out_tot_x);
E_out_tot_y_abs = abs(E_out_tot_y);

E_out_tot_x_angl = angle(E_out_tot_x);
E_out_tot_y_angl = angle(E_out_tot_y);

Vx_plot_out_tot = E_out_tot_x_abs*cos(tplot-E_out_tot_x_angl);
Vy_plot_out_tot = E_out_tot_y_abs*cos(tplot-E_out_tot_y_angl);

tmpIout = (abs(E_out_tot(1)))^2+(abs(E_out_tot(2)))^2;
%%%%%
scrn_size = get (0,'screensize');
scrn_size_x = scrn_size(3);
scrn_size_y = scrn_size(4);

%%%%%%%%%%%

%%%%%

if option_wv_pol = = 1;
W_P2_plot_option = 1;% Pol only
end

if option_wv_pol = = 21;
W_P2_plot_option = 2;% Wv only
end;

if option_wv_pol = = 22;
W_P2_plot_option = 2;% Wv only
end;

if option_wv_pol = = 31;
W_P2_plot_option = 3;% Wv only
end;

if option_wv_pol = = 32;
W_P2_plot_option = 3;% Wv only
end;

%%%%%

a_ax303a = axes('Units','Normalized','Position',[0.15,0.38,0.
27,0.54],...% [0.1,0.35,.3,.6],...
```

```
 'XTickLabel','','YTickLabel','');
set(a_ax303a,'Parent',fh303);

if W_P2_plot_option = = 1;
plot(plot_x1,plot_y1,'bv-',...
    plot_x2,plot_y2,'rs-',...
    x1,y1,'bv',x,yc1_p,'b— ',x,yc1_n,'b— ',...
    x,yc2_p,'r-',x,yc2_n,'r-',...
    Hoz_Ln_x,Hoz_Ln_y,'k:',Ver_Ln_x,Ver_Ln_y,'k:')
legend('P1','P2')
end

if W_P2_plot_option = = 2;
plot(plot_x1,plot_y1,'bv-',Voutsq_x,Voutsq_y,'ko-',...
    plot_xW,plot_yW,'ko-',...
    x1,y1,'bv',x,yc1_p,'b— ',x,yc1_n,'b— ',...
    Hoz_Ln_x,Hoz_Ln_y,'k:',Ver_Ln_x,Ver_Ln_y,'k:')
legend('P1','W')
end

if W_P2_plot_option = = 3;
plot(plot_x1,plot_y1,'bv-',Voutsq_x,Voutsq_y,'ko-',...
    plot_x2,plot_y2,'rs-',...
    plot_xW,plot_yW,'ko-',...
    x1,y1,'bv',x,yc1_p,'b— ',x,yc1_n,'b— ',...
    x,yc2_p,'r-',x,yc2_n,'r-',...
    Hoz_Ln_x,Hoz_Ln_y,'k:',Ver_Ln_x,Ver_Ln_y,'k:')
legend('P1','W','P2')
end
axis([-100 100 -100 100]);
axis('square')
title('Components')

a_ax303b = axes('Units','Normalized','Position', [0.5,0.32,.3,.6]
,...%[0.5,0.35,.3,.6],...
    'XTickLabel','','YTickLabel','');

compass(E_in(1),E_in(2),'b— ')
hold on
compass(-E_in(1),-E_in(2),'b— ')
plot(Vx_plot_out_tot,Vy_plot_out_tot,'rx-','LineWidth',2)
hold off
title('Polarization State')

text_compass_mark_a = uicontrol('Style','text',...
        'String','- - -',...
        'Units','Normalized',...
        'ForegroundColor','b',...
        'Position',[0.51,0.27,0.06,0.04],...
        'BackgroundColor',[0.925 0.914 0.847]);
```

```
text_compass_mark_b = uicontrol('Style','text',...
          'String','After P1',...
          'Units','Normalized',...
          'Position',[0.56,0.27,0.06,0.04],...
          'BackgroundColor',[0.925 0.914 0.847]);

text_compass_mark_c = uicontrol('Style','text',...
          'String','* * *',...
          'Units','Normalized',...
          'ForegroundColor','r',...
          'Position',[0.65,0.27,0.06,0.04],...
          'BackgroundColor',[0.925 0.914 0.847]);

text_compass_mark_d = uicontrol('Style','text',...
          'String','Out',...
          'Units','Normalized',...
          'Position',[0.70,0.27,0.06,0.04],...
          'BackgroundColor',[0.925 0.914 0.847]);

a_ax303c = axes('Units','Normalized','Position',
[0.9,0.3,.03,.6],...
   'XTickLabel','','YTickLabel','');
set(a_ax303c,'Parent',fh303);
bar(0,tmpIout)
axis([-0.1 0.1 0 1.1])
title('I_R_e_l');
%%%%%%%%%%%%

clear


    end
end
```

## SIMULATION D.6: IMAGING WITH A SINGLE LENS (GUI)

Simulation file: SingleLens_GUI.m
Purpose: To become familiar with image generation using a single lens
Related chapter: 6–Geometrical Optics

### FILE CONTENT

```
function SingleLens_GUI
% Single Lens
% Copyright Araz Yacoubian 2013, 2014

clear

close('all')
```

```
fh = figure('Visible','off');
set(fh,'Units','Normalized','Position',[0.05 0.43 0.44 0.42]);

set(fh,'Color',[0.925 0.914 0.847]);
set(fh,'Visible','on','Name','Single Lens','NumberTitle','off');

fwndw_H = 565;
fwndw_V = 420;
xpos2_n = 60/fwndw_H;
ypos2_n = 120/fwndw_V;
pltx_wd2_n = 450/fwndw_H;
plty_wd2_n = 250/fwndw_V;

a_ax1 = axes('Units','Normalized','Position',...
   [xpos2_n,ypos2_n,pltx_wd2_n,plty_wd2_n],'XTickLabel','','YTi
   ckLabel','');
set(a_ax1,'Parent',fh);
%%%%%%%%%%%%%%%%

%%%%%%%%%% Input:%%%%%%%%%
s1 = 50; % mm
f_mm = 25;
ha = 1;
%%%%%%%%%%%% Input:%%%%%%%%%

slider_value_s1 = s1; % mm

s1_string = num2str(s1,'%2.1f');

text_s1 = uicontrol('Style','edit',...
        'String',s1_string,'Units','Normalized',...
        'Position',[160/fwndw_H,40/fwndw_V,45/fwndw_H,25/
        fwndw_V],...
        'BackgroundColor',[1 1 1]);

text_s1_mark = uicontrol('Style','text',...
        'String','Object Distance','Units','Normalized',...
        'Position',[110/fwndw_H,30/fwndw_V,45/fwndw_H,35/
        fwndw_V],...
        'BackgroundColor',[0.925 0.914 0.847]);

sh_s1 = uicontrol(fh,'Style','slider',...
            'Max',100,'Min',0.01,'Value',slider_value_s1,...
            'SliderStep',[0.05 0.10],...
            'Units','Normalized',...
            'Position',[100/fwndw_H 70/fwndw_V 250/fwndw_H 20/
            fwndw_V],...
            'Callback',{@slider_s1_Callback});
```

```
function slider_s1_Callback(source,eventdata)
   slider_value_s1 = get(source, 'Value');
   slider_value_s1_dsp = num2str(slider_value_s1,'%2.1f');
   set(textinputSliderValue2s1, 'String', slider_value_s1_dsp);
   s1 = slider_value_s1;
   Lns_calc
end

s1_input_string = num2str(s1,'%2.1f');
textinputSliderValue2s1 = uicontrol('Style','edit',...
        'Value',s1,...
        'Units','Normalized',...
        'Position',[160/fwndw_H,40/fwndw_V,45/fwndw_H,25/
        fwndw_V],...
        'BackgroundColor',[1 1 1],...
        'String',s1_input_string,...
        'Callback',{@s1_Callback});

function s1_Callback(source,eventdata)
  s1 = str2double(get(source, 'String'));

  slider_value_s1 = s1;

     Lns_calc
end
%%%%%%%%%%%%%%%
%%%%%%%%%%%%%%%

slider_value_f_mm = f_mm;

f_mm_string = num2str(f_mm,'%2.1f');
text_f_mm = uicontrol('Style','edit',...
        'String',f_mm_string,...
        'Units','Normalized',...
        'Position',[340/fwndw_H,10/fwndw_V,45/fwndw_H,25/
        fwndw_V],...
        'BackgroundColor',[1 1 1]);

sh_f_mm = uicontrol(fh,'Style','slider',...
        'Max',100,'Min',-100,'Value',slider_value_f_mm,...
        'SliderStep',[0.0500 0.20],...
        'Units','Normalized',...
        'Position',[240/fwndw_H 40/fwndw_V 250/fwndw_H 20/
        fwndw_V],...
        'Callback',{@slider_f_mm_Callback});

function slider_f_mm_Callback(source,eventdata)
   slider_value_f_mm = get(source, 'Value');
   slider_value_f_mm_dsp = num2str(slider_value_f_
   mm,'%2.1f');%Rev 18B
```

```
    set(textinputSliderValue2f_mm, 'String', slider_value_f_mm_dsp);
    f_mm = slider_value_f_mm;
    Lns_calc

end

text_f_mm_mark = uicontrol('Style','text',...
        'String','Focal Length',...
        'Units','Normalized',...
        'Position',[380/fwndw_H,5/fwndw_V,90/fwndw_H,25/
        fwndw_V],...
        'BackgroundColor',[0.925 0.914 0.847]);

f_mm_input_string = num2str(f_mm,'%2.1f');
textinputSliderValue2f_mm = uicontrol('Style','edit',...
          'Value',f_mm,...
          'Units','Normalized',...
          'Position',[340/fwndw_H,10/fwndw_V,45/fwndw_H,25/
          fwndw_V],...
          'BackgroundColor',[1 1 1],...
          'String',f_mm_input_string,...
          'Callback',{@f_mm_Callback});

function f_mm_Callback(source,eventdata)
    f_mm = str2double(get(source, 'String'));
    slider_value_f_mm = f_mm;

          Lns_calc
end
%%%%%%%%%%%%%

   %%%%%%%%% Beg. Lns_calc%%%%%%%%

   %%%%%%%%%%%%%%%%%%%
Lns_calc

function Lns_calc
%
%
s2 = 1/((1/(f_mm))-(1/(s1)));
Mt = s2/s1; % Magnification
hb = -Mt*ha;

sys_tp = 0;
if s1 < = 0;
    tmpdispwrn = 'Keep s1 pos (>0)';
    disp(tmpdispwrn)
return
end
```

```
if f_mm > 0;
   if s2 > 0;
      sys_tp = 1;
   end

   if s2 < 0;

      sys_tp = 2;
   end

   if s2 > (10^10)*s1;
      sys_tp = 3;
      tmpdispwrn = ' Nearly Collimated. Change object
      distance or focal length';
      disp(tmpdispwrn)
      return
end

if s1 = = f_mm;
      sys_tp = 3;
      tmpdispwrn = ' Nearly Collimated. Change object
      distance or focal length';
      disp(tmpdispwrn)
      return
   end
end

if f_mm < 0;
   sys_tp = 4;
end

%%%%%%%%%%%%%%%%%%%%%%%%%%%%%%%%%%%%%%%%%%%%%%
if sys_tp = = 1;

   plot_ha_x = [-s1 -s1];
   plot_ha_y = [0 ha];
   plot_hb_x = [s2 s2];
   plot_hb_y = [0 hb];

   plot_ry1_x = [-s1 -f_mm 0 f_mm s2];
   plot_ry1_y = [ha ha ha 0 hb];

   plot_ry2_x = [-s1 0 s2];
   plot_ry2_y = [ha 0 hb];
   %%%%% Ln %%%%
   plot_Ln_x = [0 0];
   h_Ln = 1.1*max(abs(ha),abs(hb));
   plot_Ln_y_p = [0 h_Ln];
   plot_Ln_y_n = [0 -h_Ln];
   Lns_arw_top = '^';
   Lns_arw_btm = 'v';
```

```
        Linetype_img = 'b-';
        Mrkrtype_img = 'bo';
        y_min = -1.1*h_Ln;
        y_max = 1.1*h_Ln;
        %%%%% Ln %%%%
        ImageType = 'Real';

    end

    if sys_tp = = 2;

        plot_ha_x = [-s1 -s1];
        plot_ha_y = [0 ha];
        plot_hb_x = [s2 s2];
        plot_hb_y = [0 hb];

        plot_ry2_x = [-s1 0 s2];
        plot_ry2_y = [ha 0 hb];

        plot_ry1_x = [-s1 0 f_mm s2];
        plot_ry1_y = [ha ha 0 hb];

        %%%%%%%%%
        plot_Ln_x = [0 0];
        h_Ln = 1.1*max(abs(ha),abs(hb));
        plot_Ln_y_p = [0 h_Ln];
        plot_Ln_y_n = [0 -h_Ln];
        Lns_arw_top = '^';
        Lns_arw_btm = 'v';
        Linetype_img = 'r:';
        Mrkrtype_img = 'ro';
        y_min = -1.1*h_Ln;
        y_max = 1.1*h_Ln;
        %%%%%%%%%
        ImageType = 'Virtual';

    end

if sys_tp = = 4;

        plot_ha_x = [-s1 -s1];
        plot_ha_y = [0 ha];
        plot_hb_x = [s2 s2];
        plot_hb_y = [0 hb];

        plot_ry1_x = [-s1 0 -abs(f_mm) s2 abs(f_mm)];
        plot_ry1_y = [ha ha 0 hb 2*ha];

        plot_ry2_x = [-s1 0 abs(f_mm)];
        plot_ry2_y = [ha 0 f_mm*(ha/(abs(s1)))];
```

```
%%%%%%%%
plot_Ln_x = [0 0];
h_Ln = 1.1*max(abs(ha),abs(hb));
plot_Ln_y_p = [0 h_Ln];
plot_Ln_y_n = [0 -h_Ln];
Lns_arw_top = 'v';
Lns_arw_btm = '^';
Linetype_img = 'r:';%'- ';
Mrkrtype_img = 'ro';
y_min = min([-1.1*h_Ln -2*ha]);
y_max = max([1.1*h_Ln 2*ha]);
%%%%%%%%
ImageType = 'Virtual';

end

  plot_function_1

  function plot_function_1
  plot(plot_ha_x,plot_ha_y,'-',...
     'LineWidth',4)
  hold on
  plot(-s1,ha,'o','LineWidth',3)
  plot(plot_hb_x,plot_hb_y,Linetype_img,'LineWidth',4)
  plot(s2,hb,Mrkrtype_img,'LineWidth',4)
  plot(plot_ry1_x,plot_ry1_y,'- ','LineWidth',1)
  plot(plot_ry2_x,plot_ry2_y,'- ','LineWidth',1)
  plot(-f_mm,0,'o','MarkerEdgeColor','k',...
  'MarkerFaceColor','k','MarkerSize',5)
  plot(f_mm,0,'o','MarkerEdgeColor','k',...
     'MarkerFaceColor','k','MarkerSize',5)
  %%%% Ln%%%

  plot(plot_Ln_x,plot_Ln_y_p,'k-','LineWidth',3)
  plot(0,h_Ln,Lns_arw_top,'MarkerEdgeColor','k',...
     'MarkerFaceColor','k','MarkerSize',10)
  plot(plot_Ln_x,plot_Ln_y_n,'k-','LineWidth',3)
  plot(0,-h_Ln,Lns_arw_btm,'MarkerEdgeColor','k',...
     'MarkerFaceColor','k','MarkerSize',10)
     %%%%%%%%%

     x_min_a = -1.1*s1;
     x_min_b = 1.1*s2;
     x_min = min([x_min_a x_min_b 1.5.*f_mm -1.5.*f_mm]);
     x_max = max([x_min_a x_min_b 1.5.*f_mm -1.5.*f_mm]);
     %%%%%%%%%

  plot([x_min x_max],[0 0],'k-')
  %%%%%%%

  hold off
  %%%%%%%%%
```

```
            axis([x_min x_max y_min y_max]);
            grid
            tmptitle = ['f = ' num2str(f_mm) '. s1 = ' num2str(s1)
            '. s2 = ' num2str(s2) '. M = ' num2str(-Mt) '. Image
            Type: ' ImageType];
            title(tmptitle)
        %%%%%%%%%
end


%%%%%%%%%%%%%%%%%%%%%%%%%%%%%%%%%%%%%%%%%%%%%

end

%%%%%%%%%% End. Lns_calc%%%%%%%%
end
```

## SIMULATION D.7: RESOLUTION (GUI)

Simulation file: Resolution_GUI.m
Purpose: To familiarize with image degradation due to resolution limit.
Related chapter: 7–Imaging Systems

### FILE CONTENT

```
function Resolution_GUI
% Resolution 1D
% Copyright Araz Yacoubian 2013, 2014

clear

%%%%%%%%%%%%%%%%%

close('all')

fh = figure('Visible','off');
set(fh,'Units','Normalized','Position',[0.05 0.43 0.44 0.42]);

set(fh,'Color',[0.925 0.914 0.847]);
% set(fh,'Visible','on');
set(fh,'Visible','on','Name','Resolution','NumberTitle','off');

fwndw_H = 565;
fwndw_V = 420;

xpos1_n = 60/fwndw_H;
ypos1_n = 350/fwndw_V;
pltx_wd1_n = 450/fwndw_H;
plty_wd1_n = 40/fwndw_V;
```

```
xpos2_n = 60/fwndw_H;
ypos2_n = 290/fwndw_V;

pltx_wd2_n = 450/fwndw_H;
plty_wd2_n = 40/fwndw_V;

xpos3_n = 60/fwndw_H;
ypos3_n = 100/fwndw_V;
pltx_wd3_n = 450/fwndw_H;
plty_wd3_n = 150/fwndw_V;

a_ax1 = axes('Units','Normalized','Position',...
  [xpos1_n,ypos1_n,pltx_wd1_n,plty_wd1_n],'XTickLabel','','YTi
  ckLabel','');
set(a_ax1,'Parent',fh);

a_ax2 = axes('Units','Normalized','Position',...
  [xpos2_n,ypos2_n,pltx_wd2_n,plty_wd2_n],'XTickLabel','','YTi
  ckLabel','');
set(a_ax2,'Parent',fh);

a_ax3 = axes('Units','Normalized','Position',...
  [xpos3_n,ypos3_n,pltx_wd3_n,plty_wd3_n],'XTickLabel','','YTi
  ckLabel','');
set(a_ax3,'Parent',fh);
%%%%%%%%%%%%%%%%

%%%%%%%%%%%%%%%%%%%%%%%%%%%%%%

w_lpmm = 5;
slider_value_w_lpmm = w_lpmm;

w_lpmm_string = num2str(w_lpmm,'%2.1f');
text_w_lpmm = uicontrol('Style','edit',...
        'String',w_lpmm_string,'Units','Normalized',...
        'Position',[260/fwndw_H,20/fwndw_V,45/fwndw_H,25/
        fwndw_V],...
        'BackgroundColor',[1 1 1]);

text_w_lpmm_mark = uicontrol('Style','text',...
        'String','w (Lines Per mm) ','Units','Normalized',...
        'Position',[310/fwndw_H,15/fwndw_V,90/fwndw_H,25/
        fwndw_V],...
        'BackgroundColor',[0.925 0.914 0.847]);

text_w_xmm_mark = uicontrol('Style','text',...
        'String','x (mm)','Units','Normalized',...
        'Position',[380/fwndw_H,50/fwndw_V,200/fwndw_H,25/
        fwndw_V],...
        'BackgroundColor',[0.925 0.914 0.847]);
```

```
text_I_mark = uicontrol('Style','text',...
        'String','Input','Units','Normalized',...
        'Position',[10/fwndw_H,350/fwndw_V,40/fwndw_H,25/
        fwndw_V],...
        'BackgroundColor',[0.925 0.914 0.847]);

text_Iflt_mark = uicontrol('Style','text',...
        'String','Filtered','Units','Normalized',...
        'Position',[10/fwndw_H,290/fwndw_V,40/fwndw_H,25/
        fwndw_V],...
        'BackgroundColor',[0.925 0.914 0.847]);

sh_w_lpmm = uicontrol(fh,'Style','slider',...
        'Max',20,'Min',2,'Value',slider_value_w_lpmm,...
        'SliderStep',[0.10 0.10],...
        'Units','Normalized',...
        'Position',[160/fwndw_H 50/fwndw_V 250/fwndw_H 20/
        fwndw_V],...
        'Callback',{@slider_w_lpmm_Callback});

function slider_w_lpmm_Callback(source,eventdata)
  slider_value_w_lpmm = get(source, 'Value');
  slider_value_w_lpmm_dsp = num2str(slider_value_w_
  lpmm,'%2.1f');%Rev 18B
  set(textinputSliderValue2w_lpmm, 'String', slider_value_w_
  lpmm_dsp);
  w_lpmm = slider_value_w_lpmm;
  Res_Calc
end

w_lpmm_input_string = num2str(w_lpmm,'%2.1f');
textinputSliderValue2w_lpmm = uicontrol('Style','edit',...
        'Value',w_lpmm,...
        'Units','Normalized',...
        'Position',[260/fwndw_H,20/fwndw_V,45/fwndw_H,25/
        fwndw_V],...
        'BackgroundColor',[1 1 1],...
        'String',w_lpmm_input_string,...
        'Callback',{@w_lpmm_Callback});

function w_lpmm_Callback(source,eventdata)
  w_lpmm = str2double(get(source, 'String'));

  slider_value_w_lpmm = w_lpmm;

    if w_lpmm < = 20; % To avoid slider out of range error
        if w_lpmm > = 2; % To avoid slider out of range error
      sh_w_lpmm = uicontrol(fh,'Style','slider',...
        'Max',20,'Min',2,'Value',slider_value_w_lpmm,...
        'SliderStep',[0.10 0.10],...
        'Units','Normalized',...
```

```
          'Position',[160/fwndw_H 50/fwndw_V 250/fwndw_H 20/
          fwndw_V],...
          'Callback',{@slider_w_lpmm_Callback});
        end

    end

    Res_Calc
end
%%%%%%%%%%%%%%%%%%%%%%%%%%%%%%%%%%%

Res_Calc

  function Res_Calc

a_resp_LPM = w_lpmm;
%%%%

x_mm_stp = 1/1000;
x_rng_mm_hlf = 5;
x_mm = -x_rng_mm_hlf:x_mm_stp:x_rng_mm_hlf;
xfmult_mm = (1/x_mm_stp)/(2*x_rng_mm_hlf);
xf_mm = xfmult_mm.*x_mm;

[tmp,N_xmm] = size(x_mm);
fstrt_LPM = 0.7;
fend_LPM = 0.08;

tmp_T = abs((2/N_xmm).*((1:N_xmm)-(N_xmm/2)));
x_T_mm = ((fend_LPM-fstrt_LPM)).*tmp_T+fstrt_LPM;

y_mm = cos((2*pi).*x_mm./x_T_mm)+1;

yf_mm = fft(y_mm);
yf_shft_mm = fftshift(yf_mm);

tmp_h = ones(1,10);
y_mm_I = (tmp_h')*y_mm;

%%%%%%%%%%%%%
[tmp,szxfmm] = size(xf_mm);

y_flt_sq = 0.*xf_mm;
tmpsq = xf_mm-a_resp_LPM;
[tmp,sqindex] = min(abs(tmpsq));
N_cntr = (round(N_xmm/2));
sqindex_n = N_cntr-(sqindex-round(N_xmm/2));
y_flt_sq(sqindex_n:sqindex) = 1;
```

```
y_flt_sq_gsedge = y_flt_sq;
wgs_N = round(0.3*(sqindex-sqindex_n)/2);
sqindex_n02 = sqindex_n+wgs_N;
sqindex_02 = sqindex-wgs_N;

xtmp2 = sqindex_02:szxfmm;
ytmp2 = exp(-(((xtmp2-sqindex_02)./wgs_N).^2));
xtmp2n = 1:sqindex_n02;
ytmp2n = exp(-(((xtmp2n-sqindex_n02)./wgs_N).^2));
y_flt_sq_gsedge(1:sqindex_n02) = ytmp2n;
y_flt_sq_gsedge(sqindex_02:szxfmm) = ytmp2;

yf_sq_mm = yf_shft_mm.*y_flt_sq_gsedge;
y_sq_mm = ifft(yf_sq_mm);

y_sq_mm_I = (tmp_h')*y_sq_mm;
%%%%%%%%%%%%

%%%%%%%%%%%%%%%%%%
axes(a_ax1)
image(x_mm(5000:9500),tmp_h,35.*abs(y_mm_I(1:10,5000:9500)))
set(a_ax1,'XTickLabel','')
set(a_ax1,'YTickLabel','')
axes(a_ax2)
image(x_mm(5000:9500),tmp_h,35.*abs(y_sq_
mm_I(1:10,5000:9500)))
set(a_ax2,'XTickLabel','')
set(a_ax2,'YTickLabel','')
colormap(gray)
axes(a_ax3)
plot(x_mm(5000:9500),abs(y_mm(5000:9500)),'b-',...
   x_mm(5000:9500),abs(y_sq_mm(5000:9500)),'rx')
axis([0 4.5 -0.3 2.3]);
%%%%%%%%%%%%%%%%%%
   end
end
```

# Index